Perspectives on Culture and Agent-based Simulations

Studies in the Philosophy of Sociality

Volume 3

For further volumes:
http://www.springer.com/series/10961

Virginia Dignum • Frank Dignum
Editors

Perspectives on Culture and Agent-based Simulations

Integrating Cultures

Springer

Editors
Virginia Dignum
Delft University of Technology
Delft, Zuid-Holland
The Netherlands

Frank Dignum
Department of Information and Computing
Universiteit Utrecht
Utrecht, The Netherlands

ISBN 978-3-319-01951-2 ISBN 978-3-319-01952-9 (eBook)
DOI 10.1007/978-3-319-01952-9
Springer Cham Heidelberg New York Dordrecht London

Library of Congress Control Number: 2013953721

Printed on acid-free paper

Springer is part of Springer Science+Business Media (www.springer.com)

Contents

Part III Culture Simulation

Part IV Culture-Sensitive Technology Design

Chapter 1
Integrating Cultures: An Introduction

Frank Dignum and Virginia Dignum

In Sociology, the concept of formal model of culture refers to "an output from a quantitative study of collected data that seeks to describe, explain, interpret, or otherwise represent some feature, aspect, or content of culture. As a model, the output has been transformed into a summary or a representation (in reduced form) of the data that purports to be analogous (in some fashion) to the phenomena under consideration" (Mohr and Rawlings 2010). However, different disciplines in the Social Sciences take a very different approach to culture and to its influence in social behaviour. Thus it is difficult to compare and integrate the different models that are used in social science. It is also not easily possible to establish a reference model to which all the other models can be compared, because the requirements for such a reference model are very diverse, not precise and not agreed upon. Besides that the concept of culture is very abstract and vague and thus it will be impossible to give a model containing all relevant concepts (an ontology) explaining all possible relations and influences culture has on society. Therefore we advocate a more limited approach in this volume.

The aim of this volume is to analyse, from a *computational* point of view, how culture may arise, develop and evolve through time. As described in the different chapters, computational models of culture enable to represent and reason about different, possibly conflicting, social norms and practices arising from different cultures. This computational perspective enables to integrate concepts that play a role in culture but that originate from different research areas that study culture.

F. Dignum
Department of Information and Computing Sciences, Universiteit Utrecht, Princetonplein 5 De Uithof, 3584 CC, Utrecht, The Netherlands
e-mail: F.P.M.Dignum@uu.nl

V. Dignum (✉)
Faculty of Technology Policy and Management, Delft University of Technology, Jaffalaan 5, 2628 BX Delft, Zuid-Holland, The Netherlands
e-mail: m.v.dignum@tudelft.nl

V. Dignum and F. Dignum (eds.), *Perspectives on Culture and Agent-based Simulations*,
Studies in the Philosophy of Sociality 3, DOI 10.1007/978-3-319-01952-9_1,

This collection has its origin in a conference in 2010 at the Lorentz center in Leiden, the Netherlands and contains a careful selection of the papers submitted afterwards. During that meeting the role of culture in computer-based systems and virtual environments was discussed in a multi-disciplinary environment with presentations by many leading researchers on the topic from both the Social Sciences and Computer Science. Evidently, the topic is challenging. Culture is usually regarded as something vague and qualitative and perhaps the least appropriate to deal with in a computational and formal setting. Although there are some theories that make "culture" more structured and measurable, notably the famous Hofstede framework, culture in general is something that within social science is related to almost anything, for instance, religious practices, national identity, habits and customs, art and technology and social relations.

Addressing culture from a computational point of view has a twofold risk: on the one hand, the risk of reductionism, that is taking a too simplistic approach to cultural influence on behaviour; on the other hand, the risk of trying to capture too much, hence not leading to useful computational tools. The contributors to this volume are sharply aware of these risks. Their approaches and insights taken collectively show different perspectives on the potential of cultural aspects to the development of better applications and to the use of computing systems to better understand culture.

1.1 Introduction: Culture?

The term culture in its original use comes from Cicero when he talked about "cultura animi" with which he denoted the improvement of the soul. Thus the connection with the biological (or agricultural) sense of culture lays in the idea of improving (or cultivating) something. In the eighteenth and nineteenth century the term culture started to be used more in the sense that we think about it nowadays. It is seen as the range of human phenomena that cannot be attributed to genetic inheritance. In anthropology culture becomes the human capacity to classify and represent experiences with symbols and to act creatively. The symbols and results of creative activities can be material such as paintings, fashion and buildings, but also intangible such as language, music and customs.

Although the above description of the concept "culture" is still quite vague we can already see several aspects that play a role in studying culture. One of these aspects is the creative process of producing symbols to represent experiences. This usually is studied in the arts. A second aspect is the differences in culture between different parts of the world (or in general between different groups of people). What are those differences, how do they influence the people, how are they perpetuated and transmitted? These are issues related to anthropology and social science in general. It is clear that the study of culture should be an interdisciplinary study as it spans many different aspects related to how people function in a society. However, how culture is defined depends largely on the perspective that one takes and which highlights certain aspects of culture. For arts culture is related to the expression of experiences through symbols. For sociology culture might be seen as influencing social relations (like power) or as means to create a group identity.

One of the problems of precisely defining culture stems from the fact that culture in one hand is a human created phenomenon, starting from individuals, but in the other hand transcends the individuals and influences their behaviour (sometimes unconscious). This feedback of individuals to culture and back to individuals creates many complex relations and questions about purpose, origin, function and goal of culture. Moreover it makes it difficult to start with a number of basic (generally accepted) assumptions from which a general theory or model for culture can be created. Thus we will not attempt to perform this huge task within this introduction, but will limit ourselves to a more manageable pragmatic approach. We will indicate how computational models can support the comprehension of culture by creating simulations that show how societies evolve based on certain assumptions on the way culture influences behaviour. In the other hand we also need to be conscious of how our culture influences the design of computational systems. Our culture influences the way we perceive problems and their solutions with computational systems and thus which systems we create. Given these more practical issues that are studied in this volume we will in the next section describe some concepts that play an important role in our conception of culture.

1.2 What Is Culture?

Probably the most influential work trying to characterize culture is done by Hofstede (2001). Rather than actually trying to define culture in terms of other concepts this work tries to characterize culture through different dimensions of influence of culture on societies. Based on questionnaires taken in different countries about behaviour in social situations some consistent differences in the answers led to the abstraction of a number of cultural dimensions. All countries can now be given a score on each cultural dimension based on repeated questionnaires. Thus the dimensions are empirically derived. Using the cultural dimensions gives handles to explain the different type of behavioural patterns and reactions to (new) situations in different countries. Thus they form a very useful tool to study the influence of culture on societal behaviour.
The following five dimensions are distinguished by Hofstede:

- Power distance (PDI),
- Individualism (IDV),
- Masculinity/femininity (MAS),
- Uncertainty avoidance (UAI),
- Long-term orientation (LTO)

Power distance influences the expectation and importance given to power statuses. People in power positions are expected to set out directions and subordinates to obey and not take initiatives. For example: China, Russia (high PDI) opposed to Scandinavian countries (low PDI).

Individualism influences the definition of individual identity. The lower the IDV, the more one individual's identity is linked to his or her social context (e.g. relatives, colleagues). Thus, one's individual goals and actions (and the claim for this action) are more or less linked to him/herself or to his/her context. This context leads to a collective image that has to be preserved (helping each other, hiding errors, rejecting outsiders). Conversely, in high IDV cultures, individuals expect a treatment independent of any context. For example: USA, Great Britain (high IDV) opposed to South American countries (low IDV).

Masculinity indicates preferences on assertiveness, toughness, focus on performance and material success. Good performance should be recognized and rewarded, leading to competition. Conversely, low MAS cultures favour modesty, tenderness and high quality of life. Interactions focus on building cooperation and establishing consensus. For example: Scandinavian countries (low MAS) versus Japan, Italy (high MAS).

Uncertainty avoidance favours the desire for clear and explicit situations with predictable outcomes. This desire leads to establishment of rules (formal or not), making everything explicit with low ambiguity. Conversely, individuals with low UAI culture dislike the presence of rules. They tend to accept more easily situations with unspecified behaviour or unclear outcome. For example: Greece, Japan (high UAI) versus Sweden, China (low UAI).

Long-term orientation influences the time span considered when taking decisions. In high LTO culture, rewards can be sacrificed for better ones later, relationships are built on long-lasting trust and rules are flexible. Conversely, individuals in low LTO culture focus on immediate success, avoiding failure and decisions rely on dogmatic rules (e.g. total commitment, best profit commitment). For example: Extreme-Asian countries (high LTO) versus Canada, Great Britain (low LTO).

The above indicates more or less how the different cultural dimensions relate to different tendencies of behaviour. However, the dimensions do not explain why these influences work in a certain way. The closest Hofstede gets to giving a model of culture is by stating that culture is in the end based on value systems (that are shared by groups of people) and expressed in practices (rituals, norms, etc.) and symbols.

We concur that value systems can be seen as the basic drives for human behaviour (besides the biological drives). Thus in some way they will always consciously or subconsciously influence the decisions of individuals. When value systems are somehow synchronized within a group of people then their behaviour will also be similar in similar situations (note that we are not claiming that behaviour is identical, just similar). In order to facilitate the synchronization of behaviour groups will use rituals, norms and symbols that are shared between all people in the group and can thus easily be used to refer to certain value-based decisions.

Taking values and practices as the main features of culture also helps understanding how notions such as "organizational culture", "youth culture" and "subculture" relate to society culture. In each of these cases the practices that are used within the group are shared and meant to distinguish the particular group from other groups. There are also shared values, but these do not have to pertain to all facets of life, but can be confined to the part of life with which the group is concerned. Thus organizational culture is built on values that relate to how the business wants to function and be perceived. Secondly, people are not only part of an organization, but also member of other groups and the society at large. Thus they will not completely assimilate all practices and values (or only for tasks within the working context). In the case of youth culture the assimilation is bound to a certain period of life and thus also less seen as absolute. Consequence of these points is that these cultures are less pervasive and stable than society culture, but they do contain the same elements.

In the following section we will explore how culture is influencing the modelling and design of computational systems and how computational simulations can help understand culture.

1.3 Culture and Computation

The meeting of computational methods and the scientific study of culture has so far been lacking. Obviously "culture" is a theoretical term that is common to many sub-disciplines in the social sciences. However, given the lack of a common methodological framework in the social sciences, different traditions adopt a concept of "culture" that is often not compatible with the concept employed by others. Moreover, informal approaches to "culture" are unclear in their consequences and implications. Although it is not clear how a scientific approach should look like that integrated the cultural concept in the design methodology of computational systems it is clear that culture has a large impact on how systems should be designed and used. Let us just give a few anecdotal examples to make the scope of this claim clear.

Nowadays it becomes very easy to maintain and access large amounts of data locally. Within companies this leads to systems that keep track of the status of all kinds of processes and in which employees might find all kinds of information to support their task and to solve potential problems. However, storing all this information and making all this information available for all levels in the company can lead to a lot of resentment within management in high power distance cultures. In these cultures managers might feel that decisions that they are responsible for are now taken at a lower level and bypassing them, therewith threatening their status and power position. It might also lead to the fact that employees do not use the system, because they do not want to be responsible for taking the decisions. Thus the system would not render the benefits that were expected, due to the cultural biases.

In a similar way, systems that are based on argument based resolution of conflicts or systems based on individual utilities in order to coordinate tasks will function different in collectivistic and individualistic societies. Thus even the solution principles might be based on cultural biases that lead to very specific types of solutions that not necessarily work properly in other cultures. It is clear that designers should be aware of these cultural biases, especially in a world where systems are functioning world wide in many different cultures.

Of course, we could also see how computational systems help to understand culture. In particular agent based social simulation seems an interesting direction to explore the development of culture in different contexts and how different cultures can change the behaviour of a society (and e.g. its response to a new situation). Already some work has been done by Axelrod (1997) on simulations that indicate how culture disseminates and can lead to clustering behaviour. Although this work is a good starting point it illustrates nicely that the properties of culture are hidden in the functions that determine how people with different cultures mix or avoid each other. In the end culture is treated as an abstract array of features (without names) and people interact based on the similarity or difference of these arrays. Thus the fact that these arrays represent cultures is completely based on interpretation of the reader and not on any intrinsic property of the features listed in that array. On the other hand is must be said that simulations that give most insights are based on very simple principles. The interesting results come from the emerging properties from the individual interactions. Thus there is a fine balance between having too simple simulations that require immense interpretations in order to draw conclusions and very rich simulations for which so many parameters have to be set that results seem to depend on the particular parameter values. In this volume we will not solve this issue, but it includes some examples of simulations that illustrate well the state of the art and probably are starting points for further research.

When investigating the influence of culture on individual behaviour there is also possible support from computational systems. In specific agent systems seem to provide a nice basis for supporting simulations of culturally influenced behaviour. Of course, this more detailed issue has to be studied with agents that have some rules of behaviour that depend on their culture. Again one can take several approaches. One approach is to take something like the cultural dimensions of Hofstede and take the score on each dimension as the value of a parameter that influences individual decision making. It is possible to replicate culturally biased behaviour in this way by choosing the way the cultural parameters are linked to the other decision parameters carefully. However, this approach does not yet explain why people are influenced in this way. In order to answer that question more rich cognitive models are needed. These models should represent some aspects of culture and link those aspects to other factors that influence the decisions of agents (such as desires, goals, needs, etc.). There is some work done in e.g. Dechesne et al. (2013) and Mc Breen et al. (2011), but this is just the beginning of the research in this area. Some work in this volume on modelling culture could probably be used in this research, but is not in a stage yet where it could be directly implemented in the agents.

1.4 Organisation of This Volume

The chapters in this volume take different approaches to culture and can be classified along several dimensions. Together, the collection of chapters combines rigour and relevance. There is not one single simplified notion of culture, or an attempt to come to a narrow definition, but different papers address diverse aspects. Still, the coherence is high, as most of the papers focus on specific computational applications, thus demonstrating, by example, how culture can be dealt in agent systems.

Overall, this volume aims to provide an overview of the breadth and of this multidisciplinary research field, and to inspire both social and computational researchers, by describing methods, theories and concrete application results on the integration of cultural aspects into social simulations.

The first part of the volume focuses on Analysis and Modelling of Culture. The chapter "Modelling Culture with Complex, Multi-dimensional, Multi-agent Systems" by Morris, Ross, Hosseini, and Ulieru explores culture and cultural modelling from a complex systems, multi-dimensional, and multi-agency standpoint, presenting a seven-dimensional model to describe and encapsulate culture. The chapter introduces definitions, dimensions, and experiments that show the evolution and emergence of culture as a complex, distributed, social system from a computer science perspective. An extensive overview and discussion of the state of the art literature is provided. The model is used to simulate cultural interactions as a multi-agent system of high functioning agents that achieve an equilibrium of beliefs.

The chapter "Cross-Validation of Gaming Simulation and Multi-agent Simulation" by Hofstede, Jonker, and Verwaart proposes a method combining gaming and multi-agent simulation for the formulation of theories underlying trade network processes based on Hofstede's cultural dimensions (Hofstede 2001). The chapter addresses validation this type of approaches, which typically remains a problematic issue in this type of research. Two important sources of difficulties are the sensitivity of gaming simulations to the participants' cultural background and the complexity of the agent model. The proposed method enables to compare the behaviours seen in the gaming simulation with the agent-based simulation, and supports the verification that relevant sub-models of the agent-based model are valid with respect to real human behaviour.

The paper "Modelling Culture Through Social Activities" by Fuentes, Gómez-Sanz, and Pavón introduces UML-AT, a modelling language for social systems based on Activity Theory (AT). This modelling language proposes Activity Theory (Leontiev 1978) as a means to support social scientists and computer researchers to better analyse and represent the abstract requirements and computational features of social models. This framework supports the development, validation and analysis of results simulation framework. To reduce the effort in modelling, it introduces the concept of social properties as reusable specification fragments with a behavioural and organizational meaning. The use of the modelling language is exemplified by applying it to a "marital counselling" case.

The second part comprises chapters describing work on Group and Organization Culture. The chapter "Cultural Integration and Differentiation in Groups and Organisations" by Mäs, Flache, and Kitts discusses group formation and the conditions under which integration occurs. Using computational experiments, it is shown that different social forces lead to different patterns of polarization, radicalisation, and factionalism. By means of simulation, experiments are set to compare the strength and the effect of various social mechanisms, such as homophily and distancing. The focus of the chapter is on cultural homogeneity rather than cultural differences.

The paper "Modelling and Analysis of Safety Culture of Air Traffic Organizations in the National Culture Context" by Sharpanskykh and Stroeve focus on safety culture in organisations, including air traffic management, power plant control and healthcare. The authors propose an approach to systematically develop models that account for a large variety of organizational aspects, thus providing a different and structured view on safety culture from the perspective of the formal organization in relation with the variable behaviour of agents in it. The Hofstede's framework (Hofstede 2001) is applied to the problem of increasing safety in the air traffic organizations. Simulation results show how different safety measures have different effects, depending on cultural parameters.

The last paper in this part, "Monolingual Biases in Simulations of Cultural Transmission" by Roberts challenges an often-made assumption in language acquisition and uses simulation in a skilful way. It concludes that complex cultural phenomena, such as bilingualism, do not necessarily result from complex individual learning mechanisms, but much of the complexity in cultural phenomena stem from complex interactions between individuals. That is, the cultural transmission process itself can shape and influence the cultural practices it transmits.

Culture Simulation is the subject of the third part of this volume. Chapters in this part describe the use of simulation to analyse diffusion, cultural reproduction and social evolution. The paper "Towards Agent-Based Models of Cultural Dynamics: A Case of Stereotypes" by Pfau, Kashima, and Sonenberg analyse from a semi-formal perspective, the grounding model of cultural transmission, a social psychological theory that emphasizes the role of everyday joint activities in the transmission of cultural information. Their model postulates that cultural transmission happens during dialogue incidental to everyday joint activities, when interlocutors align their beliefs to a degree sufficient to carry out their joint activity. The model is based on intelligent agent research to explicate the link between agents' joint activities and the grounding process that is entailed by their task-oriented communication.

The paper "Matching and Mismatching Social Contexts" by Edmonds stresses the notion of context and context-dependence. Social Contexts are defined as specific types of recognised social situation for which specific norms, habits, rules, etc. are developed over time. The author explores the implications of social context to the problem of integrating cultures. The mapping of social contexts in different cultures greatly influences both the outcomes of meeting cultures and the steps that might be taken to facilitate their integration. That is culture is structured in a fundamental manner by social context. The chapter "The Role of Stability in Cultural Evolution: Innovation and Conformity in Implicit Knowledge Discovery" by Bryson discusses the role and sources of innovation in generating culture, and also the role of norms in preserving it. It presents a conceptual approach relevant to culture research, as it explicitly addresses theoretical controversies in this area. In this chapter, simulation is used to test and discriminate hypotheses about social/cultural phenomena, rather than seeking applications.

Finally, the forth part focuses on Culture-Sensitive Technology Design. In the chapter "Socially-Oriented Requirements Engineering–Software Engineering Meets Ethnography" by Pedell, Miller, Vetere, Sterling, and Howard, a different view on culture is presented, that of social practice. The chapter outlines an approach for eliciting, understanding, and representing the cultural aspects of the domestic environment for the purpose of system design by using agent models as shared artefacts to represent the everyday cultural life of the home. Conclusions show that the approach described can assist ethnographers and software engineers in arriving at a shared understanding of social goals and the related interactions in a way that became useful in ongoing software development for the social domain.

The chapter "Cultural Broker Agents: A Framework for Managing Cultural Misunderstandings" by Gonzáles, Barthés, and Ramos presents a system intended at reducing the impacts of cultural differences in multi-cultural collaboration based on constructing quantifiable cultural profiles. The work is applied to the identification and management of conflicts in communication. The authors propose a platform based on cognitive agents for improving multicultural interactions. Agents manage cultural profiles and obtain contextual information about user interactions. The framework is based on a formalization of the Hofstede framework (Hofstede 2001) based on fuzzy logic.

The chapter "Culture Driven Game Design Method: Adapting Serious Games to the Players' culture" by Meershoek, Kortmann, Meijer, Subrahmanian, and Verbraeck clearly shows the relevance of culture in the design of serious games. When culture is not incorporated in the game design phase, it may still be discovered during the testing, but this may be too late. The paper suggests how this can be done differently and more efficiently.

1.5 Discussion

The contributions in this volume are a valuable contribution towards the understanding of culture and its relation to computational systems. The interdisciplinary nature of culture is reflected in the contributions which come from diverse disciplines and highlight different aspects of culture. They show how culture can be modelled from different perspectives, but also how culture influences models for computational systems. Given the complexity of culture we cannot hope to cover all aspects of culture or give a definite answer on the relation between culture, society, individuals and the computational systems that they use. We do believe, however, that this volume is a good starting point for research on integrating cultures and computational systems. Thus it can be seen as the start of an interdisciplinary dialogue on culture where connections between the different perspectives are discovered and which forms a basis of understanding and proper use of culture in the systems that we build.

References

Axelrod, R. (1997). The Dissemination of Culture: A Model with Local Convergence and Global Polarization. *Journal of Conflict Resolution*, 41(2):203–226.

Dechesne, F., Tosto, G. D., Dignum, V., and Dignum, F. (2013). No smoking here: values, norms and culture in multi-agent systems. *Artificial Intelligence and Law*.

Hofstede, G. (2001). *Culture's consequences: Comparing values, behaviors, institutions, and organizations across nations*. Sage Publications.

Leontiev, A. N. (1978). *Activity, Consciousness, and Personality*. Prentice-Hall.

Mc Breen, J., Di Tosto, G., Dignum, F., and Hofstede, G. J. (2011). Linking Norms and Culture. In *Second International Conference on Culture and Computing*, pages 9–14. IEEE.

Mohr, J. W. and Rawlings, C. (2010). Formal models of culture. In Hall, J., Grindstaff, L., and cheng Lo, M., editors, *Handbook of Cultural Sociology*, pages 118–128. Routledge.

Part I
Analysis and Modelling of Culture

Chapter 2
Modelling Culture with Complex, Multi-dimensional, Multi-agent Systems

Alexis Morris, William Ross, Hadi Hosseini, and Mihaela Ulieru

2.1 Introduction: Modelling Organizational Cultures

No single definition of a social science construct is likely to do justice to its complexity.
—Hofstede (2001)

This chapter focuses on a new approach to model and discuss culture and explores the emergence and evolution of culture within organizations. This is a first step toward future studies on the interplay and eventual integration of different cultures in a shared environment. The primary theme throughout this work is that in order to understand, discuss, and measure culture, it must be recognized as a complex, multi-dimensional, and multi-agent system. These three aspects are the proposed foundation for experiments in culture beginning at the level of the individual unit and progressing toward how groups of such units form and influence a cultural system.

Culture plays a key role in organizations, both as a determinant of relationships among individual units of the organization and as a macro-level driver of its behaviour. It should be considered as one of the main points of analysis when modelling organizations (see Hofstede 2001, Chap. 8, for more on culture as it relates to organizations). Cultural modelling allows for incorporating knowledge

A. Morris (✉) • W. Ross • M. Ulieru
Faculty of Computer Science, Adaptive Risk Management Laboratory, The University of New Brunswick, Fredericton, NB, Canada
e-mail: alexis.morris@unb.ca; william.ross@unb.ca; ulieru@unb.ca

H. Hosseini
Faculty of Computer Science, Adaptive Risk Management Laboratory, The University of New Brunswick, Fredericton, NB, Canada
School of Computer Science, University of Waterloo, Waterloo, ON, Canada
e-mail: h5hosseini@uwaterloo.ca

V. Dignum and F. Dignum (eds.), *Perspectives on Culture and Agent-based Simulations*, Studies in the Philosophy of Sociality 3, DOI 10.1007/978-3-319-01952-9_2,

about the effect and influence of culture on an organization and predicting how the type of culture at work affects the ability of the organization to function, achieve its goals, and ultimately survive.

In order to adequately model and simulate organizational cultures, there are four key components explored in this work: first, a fitting and tangible definition of culture is required; second, a study of the key dimensions of culture is necessary; third, these key dimensions must be used to establish cultural parameters; and, finally, a method of simulating the organization with the defined cultural parameters is needed. Together, these provide the methodology, tools, and techniques for setting up and conducting experiments involving culture in organizations.

Contributions of this chapter are three-fold: (i) it adds to the literature of culture as a complex system, (ii) it presents a new seven-dimensional model to describe and encapsulate culture, and (iii) it models cultural interactions as a multi-agent system of high functioning agents that achieve a certain equilibrium in beliefs. These are elaborated further in the chapter: Sect. 2.2 discusses organizational modelling and presents a working definition of culture; Sect. 2.3 describes the notions behind a complex system and makes the case for culture as such a system; Sect. 2.4 proposes a new model for culture using seven dimensions and provides the reasoning behind this approach; Sect. 2.5 discusses relevant literature regarding culture models; Sect. 2.6 describes how to measure culture with high-functioning agents; Sect. 2.7 explores both the emergence and evolution of culture and discusses the experimental results; and Sect. 2.8 concludes the chapter.

2.2 Organizational Modelling and Culture

An organization is defined as a social arrangement which pursues collective goals, controls its own performance, and has a boundary separating it from its environment (Alvesson 2003; Hatch and Cunliffe 1997). As such, organizational models must account for not only the individual units, but also for the behaviour and interaction patterns of these units, which at a higher meta-level can be seen and described as a culture. Such models are useful in simulations of real-world organizations under a host of conditions, allowing for large volumes of experiments to be conducted in a controlled environment. To perform similar experiments in an in-vivo fashion would be expensive. The results from such studies allow for detailed analyses that can be useful in predicting organizational states and behaviours. This predictive capacity helps in translating simulation knowledge directly into the real world through targetted policy-making and best-practices based on the model.

Cultures are unique to organizations, based on the complex relationships between the parts of the organization and other factors such as environment or technologies (see Ashkanasy et al. 2000, Chap. 6, for more on how key relationships develop meaning and culture). These relationships at lower levels diversify organizations from each other in important and unique ways that can be compatible, complementary, or competitive. The effects of such relationships are seen in varying degrees within all systems, especially when considering the unique interplay between systems of systems, including human societies.

2.2.1 A Working Definition of Culture

Traditionally, culture is defined as a "set of shared attitudes, values, goals, and practices that [both] characterizes an institution, organization, or group" and emerges from and sets the behaviour of a group (Kroeber et al. 1952). It has also been considered by social scientists to be the "collective programming of the mind" (Hofstede 2001, Chap. 1). In Ashkanasy et al. (2000, Chap. 10), three perspectives of culture are defined: the integration perspective, where people share a common set of beliefs; the differentiation perspective, where different subgroups have different beliefs, but must learn to resolve conflict; and the fragmentation perspective, where, because of such ambiguity in beliefs, individuals continually fragment into ever-changing subgroups. In this work, it is the integration perspective that is being adopted, as well as the view that culture is an open system in a state of equilibrium (Von Bertalanffy 1968).

Our unique working definition of culture is proposed as *the holistic interaction among n agents, across seven distinct dimensions, that results in the stabilization of beliefs within these interacting agents over time*. This allows us to consider both the community of individuals as a whole (e.g. a country or an institution), as well as distinct parts (e.g. a province or a department) with their particular characteristics. This general definition can extend from a single, mono-cultural context to a more diversified, multi-cultural one. At the same time, it frames "culture" as a multi-agent system.

2.3 Culture as a Complex System

This section promotes the view of culture as a complex system, and makes the case that complex systems theory provides strong tools to capture and delineate culture. Culture has been studied in many works and contexts over a wide range of literature domains, and may be considered as one of the "fuzzy" human-factors which are well known, but largely intangible. The view of culture as a system promotes a focus on the emergence of culture from its tangible components, and how the relationships between these components openly affect the meta-level culture, and how the culture, in turn, affects these components.

A complex system may be understood from "the amount of information needed in order to fully describe the system" Bar-Yam (1997, Chap. 8). This includes information about the system states and component interactions at all levels (or scales) of the system, from high-level to low-level. For culture, the system components are as follows. Elements are individuals within a system that are autonomous and belief-based. Interactions between these are seen as social communication, both verbal (spoken or written) and non-verbal (social or emotional cues, or levels of influence) channels. Other complex systems concepts like reproduction, growth, and feeding are also relevant, at the low-level (Bar-Yam 1997). Culture *reproduces* as the spread of beliefs from one system achieves stabilization within another system; culture *grows* as more individuals adopt/share the same beliefs; and culture *feeds* (or is strengthened) as beliefs are reinforced and become more resilient to change. The main complex systems concepts in this chapter are (a) emergence, (b) evolution, and (c) equilibrium. Emergence is the notion that "the whole is more than the sum of parts. . .that constitutive characteristics are not explainable from the characteristics of isolated parts. . .[but] appear as 'new' or 'emergent'" (Von Bertalanffy 1968, Chap. 3). Hence culture, once it has emerged, is something more than its elements. Evolution may be considered as the accumulation and advancement of high-level changes in a system over a period of time (Von Bertalanffy 1968). This accumulation of changes may occur across any significant property of the system, in any direction, as trends. In terms of culture, evolution is seen as the global trends of beliefs changing in both its high-level and low-level elements, across any of its dimensions over time. Finally, equilibrium is the balance, or "centeredness" within a system (Von Bertalanffy 1968), that stems not only from the interactions within the system, but also from the strength of those interactions. This equilibrium emerges from the lowest levels of the system. These, in conjunction with the factors mentioned above, can provide a strong ontology for discussing culture from the complex systems standpoint.

2.4 A Multi-dimensional Framework for Culture Modelling

Modelling culture requires a broad perspective that is capable of capturing its complexity while still being concrete enough for simulations. We propose an approach involving seven dimensions of culture for organizations. These extend upon our previous work on organizational modelling (Bicocchi et al. 2010) and include the physical, individual, functional, structural, social, normative, and information dimensions. These seven dimensions, each described below, provide a new way to discuss culture and its parameters. It should be noted that some factors appear in more than one dimension. This speaks to the interconnectedness of dimensions.

2.4.1 Physical

The *Physical* dimension of culture relates to its components in the actual world, ranging from the tools and technology in use, to the forms of its common assets (e.g., buildings, cars, and clothing). In every organizational system, environmental aspects such as size, location, physical distance, and quality of life affect the behaviour of agents within that system. Additionally, physical characteristics of the agents themselves are also important. For example, size and gender can play an important role in forming cultures.

2.4.2 Individual

The *Individual* dimension describes the component actors in the system and elucidates their unique characteristics, which eventually propagate throughout the culture. Individual factors, both physical and cognitive, highly affect a culture. Cognitive elements are beliefs and desires built up over time that form innate personality, degree of conformity, interests, and experiences. Other attributes are acquired by social interactions and what influential third parties (authorities or experts) believe. At this level, local and personal values are widely expressed within the organization and behaviour can be studied. These elements modify the attributes within the members and can influence the evolution of culture.

2.4.3 Functional

The *Functional* dimension associates a particular role to the individuals within the system, dictating their permissible actions. Similar functions between individuals encourage closer associations and group formations. For instance, medical-related professions such as doctors and nurses develop a similar culture to interact within their organizations. They share (some) knowledge about their domain and communicate through a known ontology. Such functional diversity influences the cultural cohesiveness among groups of individuals.

2.4.4 Social

The *Social* dimension is used to classify the type of interaction that takes place between system actors (e.g., the particular nature and medium of social communication) and the frequency of this interaction. It also refers to specific properties of the relationship between individuals, such as trust and reputation. This dimension determines the kind of social network that unfolds within the system and how resilient that network is to change and, in turn, how resilient the culture is to new beliefs.

2.4.5 *Structural*

The *Structural* dimension of culture characterizes the formal organizational network that exists within the system. Traditional organizations shape their structure based on hierarchical levels of authority (e.g., chain-of-command of superiors, subordinates, and colleagues). This not only affects the culture between different levels of the hierarchy, but also promotes the formation of sub-cultures. The form of the structure changes the behaviours, norms, and understanding of members and, in this way, affects the culture.

2.4.6 *Normative*

The *Normative* dimension characterizes policies and rules that govern the behaviour of individuals within a culture. These may evolve in a bottom-up manner (Hosseini 2010; Savarimuthu 2007) and can be formal, written for a certain environment, or informal, based on descriptive actions of the members of the organization and traditions. Culture emerges from the aggregation of norms that are common to a group of agents (Dignum and Dignum 2009) and can impact decision making and the degree of autonomy among individual agents (Conte et al. 1988; Dignum et al. 2009).

2.4.7 *Information*

The *Information* dimension represents the type, speed, and content of information elements used by individuals in the system. Information has many meanings as a concept (Floridi 2002) and is closely related to notions of communication, control, data, knowledge, meaning, pattern, and representation. This is seen in modern cultures where information exchange is facilitated by technological advancements that allow for swifter adoption of ideas, and hence more dynamic cultures.

2.5 Related Work on Cultural Modelling

Approaches to modelling culture from a multi-dimensional perspective are not new. Other key dimensions have been identified in organizational culture literature as seen in Ashkanasy et al. (2000). Hofstede (Chap. 25), for instance, promotes a four-dimensional and a six-dimensional model. The four-dimensional model targets culture as it relates to nations and governments, while the six-dimensional model targets organizations. Payne (Chap. 10), presents a three-dimensional model of culture; Ashkanasy et al. (Chap. 8), promote a ten-dimensional model of culture; and Dickson (Chap. 28), presents a nine-dimensional model. These are seen in Table 2.1, alongside the framework presented in this chapter.

A detailed comparison between these models is left for future studies. However, the primary difference is that the seven-dimensional model has been designed with multi-agent systems simulations in mind and is a more general ontology. The approach targets a description of an organizational culture that can be built into properties of individual agents and encourages a holistic approach to modelling culture. In many ways, the approach of the seven-dimensional model for agents is generic and, arguably, subsumes the other multi-dimensional models. For instance, both Hofstede's "power distance" and Payne's "strength of consensus" dimensions could be included as factors within the social dimension.

This chapter focuses primarily on the bottom-up interactions of the cultural system and, as such, uses an agent-based modelling approach. The reader is referred to our previous work in Morris et al. (2011) and Hosseini and Ulieru (2011) for other related aspects of culture modelling involving agent-based interaction models, norm-governed models, learning and adaptation in cultures, and mathematical techniques, in addition to multi-dimensional descriptions of culture.

2.6 Modelling and Simulating Organizational Culture in a Multi-agent System

From our definition, culture represents a shared understanding of a set of beliefs that determines, among other things, accepted behaviour (Kroeber et al. 1963). The way in which culture emerges is based heavily on members of the organization. Particularly, the position taken in this chapter is that the influence of existing organizational members affects the culture of new members. While each member of the organization may have his or her own particular beliefs about a specific element, ultimately there is an overarching belief that becomes dominant in the culture. In this section, the mechanisms used to store cultural beliefs (i.e., the cultural belief set), calculate influence, and modify beliefs for each agent will be examined.

Literature to support these mechanisms is found in Ashkanasy et al. (2000). For example, in Ashkanasy et al. (2000, Chap. 3), the emergence of culture results from social actors engaging in processes called "events." Anyone participating in an organization does so by interpreting events and influencing the meanings that others give to them. Powerful organizational actors, such as managers, are able to create meaningfulness for other agents through formal or informal organizational rules (or norms). These develop and change through the actions of numerous actors as they establish, enact, enforce, misunderstand, resist, and/or break the rules (Ashkanasy et al. 2000, Chap. 6). Culture is determined precisely by the configuration of the rules and actors involved. Various influence models have also been discussed in the literature, and influence factors include role (e.g., superior, subordinate, and

Table 2.1 Multi-dimensional culture models found in literature and the proposed seven-dimensional model

Hofstede model for nations (1990)	Hofstede model for organizations (1990)	Payne model (1996)	GLOBE model for organizations (1999)	Ashkanasy organizational culture profile (2000)	Seven dimension model for agents (2010)
Power distance	Process-oriented vs. Results-oriented	Strength of consensus	Power distance	Leadership	Physical
Uncertainty avoidance	Job-oriented vs. Employee-oriented	Pervasiveness	Uncertainty avoidance	Structure	Structural
Individualism vs. Collectivism	Professional vs. Parochial	Psychological intensity	Humane orientation	Innovation	Functional
Masculinity vs. Femininity	Open-system vs. Closed-system		Assertiveness	Job performance	Individual
Long-term orientation vs. Short-term orientation	Tightly controlled vs. Loosely controlled		Gender egalitarianism	Planning	Social
	Pragmatic vs. Normative		Future orientation	Communication	Normative
			Performance orientation	Environment	Information
			Individualism vs. Collectivism	Humanistic workplace	
			Organizational collectivism	Development of the individual	
				Socialization on entry	

colleague), self, and leadership characteristics of the individual (Ashkanasy et al. 2000, Chaps. 6, 10). These have been captured already, along with other factors, using our seven-dimensional modelling approach.

2.6.1 Cultural Belief Set

The cultural belief set (CBS) contains beliefs that exist in the organization's cultural landscape. These may be beliefs about particular attitudes, values, goals, or practices. Each belief in the CBS can assume one of three values, based on deontic logic: prohibited, permitted, or obligated. As an example, a belief that "punctuality = prohibited" means that it is culturally unacceptable to be punctual; "punctuality = permitted" means that it is culturally neutral whether or not someone is punctual; and "punctuality = obliged" means that it is culturally required to be punctual.

Since the belief value in the CBS has been restricted to three possibilities, the current culture's stance on a particular cultural belief, x, in the CBS can be ascertained by determining which of the three possible values has the greatest consensus among the various members of the organization.

2.6.2 Influence Calculation

The influence of one agent over another agent is used as the mechanism for changing culture. It is based on the notion described previously that key individuals in the organization have a greater influence on its culture. This influence can be computed using factors from each of the seven dimensions. The factors in Table 2.2 have been incorporated into the influence calculation and are part of the influence factor set (IFS).

The equation used to calculate the influence of one agent over another is presented in Eq. 2.1. The IFS factors have been included, along with an impact ratio, α_j, for each factor. The latter allows the particular factor's influence to be customized for each agent.

$$\iota_1 = \sum_{j=1}^{p} (IFS_a(j) - IFS_b(j)) * \alpha_a(j), \tag{2.1}$$

where p is the number of items in the influence factor set (IFS) involving $agent_a$'s beliefs about $agent_b$ (i.e., items 1–7 in Table 2.3); j is an index to a row in the IFS table and α is the corresponding impact factor; IFS_a and IFS_b are the influence factor sets for $agent_a$ and $agent_b$, respectively.

Table 2.2 Factors incorporated into the influence calculation and influence factor set (IFS)

Cultural influence factors		
Structural	1	How does agent A relate structurally (within the context of an organization) to agent B? {supervisor, subordinate, colleague}
Physical	2	How close is agent A's workstation from agent B's workstation? {proximity_Threshold} (agent A has a greater chance of being influenced by agents within its proximity threshold)
Functional	3	How similar is agent A's role to agent B's role? [0–1]
Individual	4	Do agent A and B share the same gender? {true, false} (agent A has a greater chance of being influenced by an agent with the same gender)
	5	Are agent A's and B's personalities congruent? [0–1] (agent A has a greater chance of being influenced by an agent with a congruent personality)
	6	How does agent A's experience in the organization compare with agent B's experience? (agent A has a greater chance of being influenced by an agent with more experience)
	7	How does agent A's leadership ability compare with agent B's leadership ability? (agent A has a greater chance of being influenced by an agent with more leadership ability)
Normative	8	Is the particular belief from the CBS formally or informally specified? (an agent has a greater chance of quickly shifting its cultural belief if it relates to a norm that is formally specified)
Social	9	Does agent A seek peer validation from agent B? [0–1] (this may be due to several factors)
	10	Does agent A trust agent B? [0–1]
Information	11	Through what medium does agent B principally communicate to agent A? {face-to-face > Web 2.0 > phone > email}
	12	Does agent A experience the cultural feedback first-hand or second-hand from agent B? (this speaks to the strength of the confidence interval)
	13	If directly, does agent A receive feedback via verbal or non-verbal cues? (this speaks to the strength of the confidence interval; besides verbal cues may be misinterpreted)

Equation 2.2 represents a similar calculation, but for internal influences (e.g., preferences) of $agent_a$ that do not involve $agent_b$ directly.

$$\iota_2 = \sum_{j=p+1}^{n} IFS_a(j) * \alpha_a(j), \tag{2.2}$$

where $p + 1$ is the first item of the IFS that does not involve $agent_b$; n is the total number of items in the influence factor set (i.e., items 8–13 in Table 2.3); j is an index to a row in the IFS table and α is the corresponding impact factor.

The total influence calculation for $agent_a$ is seen in Eq. 2.3.

$$\iota_a = \iota_1 + \iota_2 \tag{2.3}$$

Table 2.3 Influence and impact factors used in the CBS (α values assigned in simulation)

Item no.	Influence factors	Impact ratios (α)
External influences		
1	Structural relation	Structural impact ratio
2	Workstation proximity	Distance impact ratio
3	Role similarity	Role impact ratio
4	Gender	Gender impact ratio
5	Personality similarity	Personality impact ratio
6	Experience similarity	Experience impact ratio
7	Leadership similarity	Leadership impact ratio
Internal influences		
8	Formally specified	Formality impact ratio
9	Seek validation	Validation impact ratio
10	Trust	Trust impact ratio
11	Communication medium	Communication impact ratio
12	First-hand feedback	First-hand impact ratio
13	Verbal feedback	Verbal impact ratio

2.6.3 *Updating the Cultural Belief Set*

In the simulation, agents share cultural beliefs with other agents whenever a cultural event takes place. These events occur whenever an agent tests a cultural belief in its CBS'. (CBS' is used to distinguish the agent's personal belief set from the organizational belief set, CBS.) These events take the form of a fact in the world, e.g., $agent_a culturalbelief = value$. The current agent, $agent_a$, is enacting a specific belief in its CBS'. This agent will receive direct feedback—praise or chastisement—from the other agents in the organization. This feedback is in the form of $agent_b culturalbelief = value$. If the value from $agent_b$ matches $agent_a$'s value, the behaviour or belief is being positively reinforced; otherwise, it is being negatively reinforced.

An agent's cultural beliefs are reconsidered everytime the agent experiences an event. The other agents also experience the event, but their feedback is received second-hand, or indirectly. Events that are experienced first-hand by the agent will have a greater impact on the value of a cultural belief than events that are experienced second-hand. This is accomplished via $IFS(12)$ in Table 2.3.

For each belief, x, in an agent's CBS', a confidence value is associated with each of the three possible values—i.e., prohibited, permitted, or obliged. In order for the value of x to change, the confidence related to one of the other possible values must become the new maximum. These confidence values are based on the beliefs expressed by other agents, following a cultural event, combined with the influence of other agents based on previous calculations in Eqs. 2.1–2.3. For instance, dressing casually may start as a prohibited belief for $agent_a$, but as more and more interactions take place with different belief values, eventually the permitted or obligated value may become the new maximum, meaning that $agent_a$'s belief value will change. Equations 2.4–2.6 show the confidence calculations associated with the three possible values of belief x inside $agent_a$'s CBS'.

$$\Phi_{prohibited}(x) = \sum_{i=1}^{k} \frac{\beta(x, i, prohibited) * \iota_i}{k}, \tag{2.4}$$

$$\Phi_{permitted}(x) = \sum_{i=1}^{k} \frac{\beta(x, i, permitted) * \iota_i}{k}, \tag{2.5}$$

$$\Phi_{obligated}(x) = \sum_{i=1}^{k} \frac{\beta(x, i, obligated) * \iota_i}{k}, \tag{2.6}$$

where x is the belief under consideration in the CBS'; k is the number of agents in the system; ι_i is the influence of $agent_i$ on the current agent (in Eq. 2.3); β is the function below which produces a 1 if $agent_i$'s value for belief x matches the value currently under consideration, i.e., μ, which is one of the three possible values of x: prohibited, permitted, obligated.

$$\beta(x, i, \mu) = \begin{cases} 1 \text{ if } CBS'_i(x) = \mu \\ 0 \text{ otherwise} \end{cases} \tag{2.7}$$

After each cultural event, the agents recompute confidence for all three possible values of each belief in their CBS'. Ultimately, the belief value with the greatest confidence will be selected by the agent as cultural belief x. However, if an agent's confidence is below a certain threshold (unique to the agent), then the agent will feel free to "test" this cultural belief with counter-cultural behaviours, i.e., the agent may perform an action that is counter to the belief value in the CBS. Such "agents-of-change" (Ulieru and Verdon 2009), if combined with high influence, may eventually shift an organization's CBS into a new equilibrium.

2.7 Experiments

The previous section outlined the foundations used to develop our culture simulation, and in this section we test these notions in a simple, hypothetical organization (its roles and structure) using multi-agent techniques. We model a set of workers, having unique individual characteristics and roles. We have chosen to use the Brahms multi-agent development environment (Clancey et al. 1998) that builds on

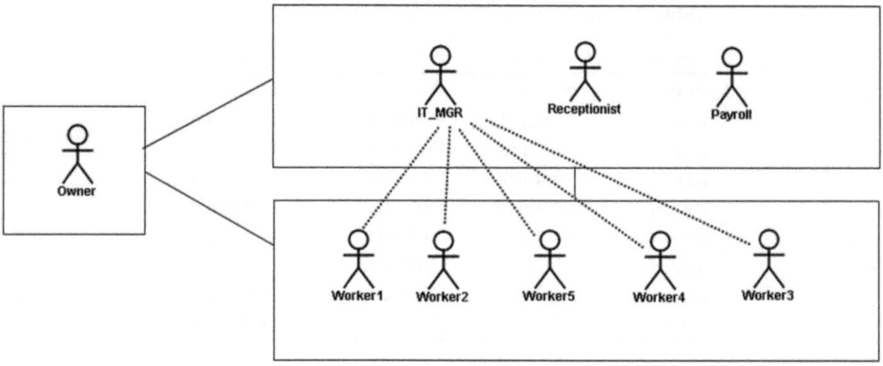

Fig. 2.1 A simple example of an organization consisting of nine agents. The most influential agents are the owner, IT manager, and payroll manager. Each agent is fully connected with all other agents. The *dotted lines* indicate supervisor-subordinate relationship between the IT manager and worker agents

the Beliefs-Desires-Intentions (BDI) paradigm (Rao and Georgeff 1995), with the concept of work practice, which attempts to capture what workers actually do in a typical day (as opposed to what workers should do).

2.7.1 Scenario

In our experiments, a small generic organization is considered, along with the following roles: an owner ($agent_1$), IT manager ($agent_2$), receptionist ($agent_3$), payroll manager ($agent_4$), and five generic worker agents reporting to the IT manager ($agents_{5-9}$), as seen in Fig. 2.1. The CBS is comprised of the following elements: (i) working after hours (overtime), (ii) appropriate business attire, and (iii) punctuality; and the culture of the organization can be determined at any given time based on the majority consensus of whether these beliefs are prohibited, permitted, or obligated. Each agent is instantiated with an initial set of beliefs pertaining to the CBS, as seen in Table 2.4, in addition to initial influence factors and impact ratios which were described previously. Agents in the organization are fully connected to each other in this scenario, having 'subordinate-to' and 'colleague-of' relationships based on role. Future experiments can explore different network configurations to see their effects on culture, but a fully-connected case is presented here as a first step.

In order to show emerging culture, we demonstrate how the belief set equilibrium of our basic organization is affected under three conditions: (i) the effect of adding the most influential agents at the beginning, (ii) the effect of adding the most influential agents in the middle, and (iii) the effect of adding the most influential agents at the end. The addition of an agent may shift the equilibrium of the

Table 2.4 Initial values for each agent's CBS'

Agent	Overtime	Formal attire	Punctuality
$agent_1$	Permitted	Prohibited	Obligated
$agent_2$	Obligated	Prohibited	Obligated
$agent_3$	Obligated	Prohibited	Permitted
$agent_4$	Prohibited	Obligated	Permitted
$agent_5$	Prohibited	Obligated	Obligated
$agent_6$	Prohibited	Obligated	Permitted
$agent_7$	Obligated	Obligated	Prohibited
$agent_8$	Prohibited	Obligated	Permitted
$agent_9$	Obligated	Prohibited	Obligated

organization's culture, as each agent will have a different cultural influencing factor dependent on such things as role occupied, personality, and existing social connections within the organization.

2.7.2 *Visualizing the Cultural Belief Set*

By modelling each agent individually, each can have its own unique beliefs about culture. When multiple agents begin interacting, certain forces will cause some beliefs to be accepted by the community and become part of the culture (i.e., part of the social memory). Such a force may be a new manager, for example, who has authority over particular agents. Moreover, we believe that culture stabilizes as more agents join the organization, so it becomes resilient to change. However, we still maintain that if a major destabilizing force occurs (e.g., a key agent such as a manager in an organization is replaced), then a cultural shift may occur, resulting in a new equilibrium. To display culture, we use the notion of a *belief set equilibrium*, which represents changes in beliefs over all agents in the system.

This equilibrium is seen in the experiments below, represented as radar plots. The size of the plot indicates the number of agents in the system, or how mature the culture is. The shape of the plot indicates the orientation of the cultural system. Finally, the time-steps show the progression of the culture from a small organization of three agents to a larger group, and the variation between time-steps represents the cultural evolution in the system.

2.7.3 *Experiment 1: Adding the Most Influential Agents at the Beginning*

In this experiment, the organization begins with the three most influential agents: the owner and the two managers. These agents then have 1 simulated month to

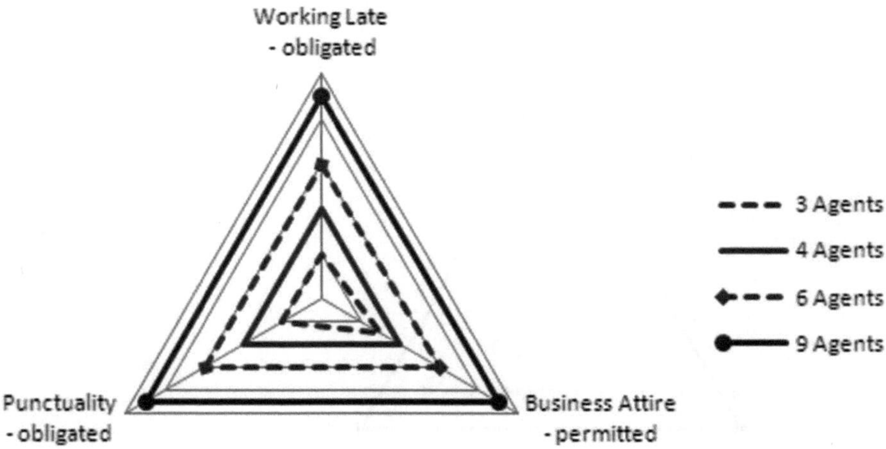

Fig. 2.2 Experiment 1: Adding most influential agents at the beginning. Cultural beliefs stabilize after the fourth agent is added

perform cultural interactions. During this time, for this experiment, two of the agents agree that employees must work after hours and be punctual, and all three agree that business attire is not that important. After the 1 month period, another agent is added to the organization. Once again, the agents have a month to perform cultural interactions before the next agent is added.

As can be seen in Fig. 2.2, once four agents are added to the organization, the cultural belief set stabilizes and other agents added to the system adopt the organization's culture. This is because the existing agents are sufficiently influential and eventually convince all other agents within the organization to conform to their culture.

2.7.4 Experiment 2: Adding the Most Influential Agents in the Middle

In this experiment, the organization's three most influential agents are added after three other agents perform cultural interactions for a month. As in the previous experiment, the additional agents are added subsequently after a 1-month simulated period. This continues until all nine agents have been added to the organization.

As can be seen in Fig. 2.3, complete stabilization of the culture does not occur until six agents have been added to the organization. This suggests that the influence of the most powerful agents impacted the initial culture of the organization, which existed during the first month when three initial agents were present. This likely occurred because none of the first three agents were sufficiently influential to convince the others to adopt their cultural position.

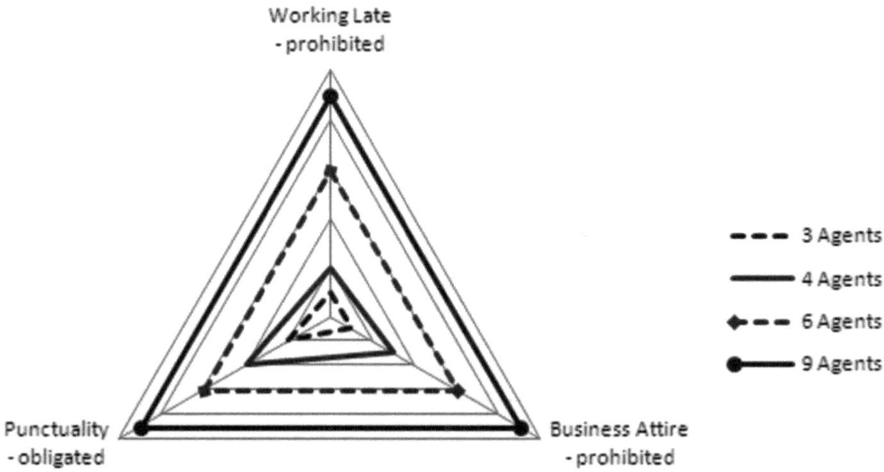

Fig. 2.3 Experiment 2: Adding most influential agents in the middle. Cultural beliefs stabilize after the sixth agent is added

2.7.5 Experiment 3: Adding the Most Influence Agents at the End

In this experiment, the organization's three most influential agents are added to the organization as the last three agents. Once again, they are added in monthly increments, following the initial three agents and the three subsequently added lesser influence agents. This particular experiment may simulate the case where some key management is replaced at some interval during the lifetime of the organization.

As can be seen in Fig. 2.4, complete stabilization of the culture occurs once six agents have been added to the organization. This suggests that even though the most influential agents are not added until the end, the first six agents are able to create enough "pull" together to compensate for the greater influence of these other three agents. Because these influential agents are added individually, neither one alone is able to overcome the cultural stability already existent within the organization.

2.8 Conclusion

Culture is not only an intangible social construct, but also an emergent property, and the primary theme of this chapter is that in order to understand, discuss, and measure culture it must be recognized as a complex, multi-dimensional, and multi-agent system. In this work, culture has been defined and considered holistically, from both a top-down and bottom-up perspective.

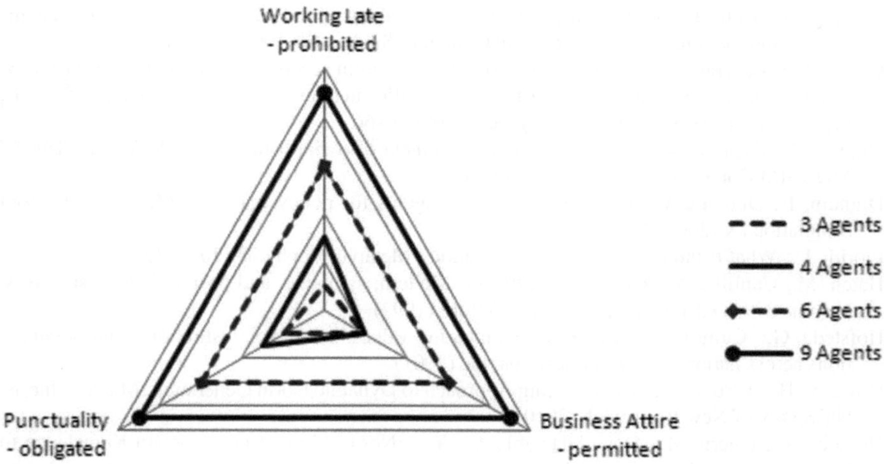

Fig. 2.4 Experiment 3: Adding most influential agents at the end. Cultural beliefs stabilize after the sixth agent is added

The multi-dimensional modelling work in this chapter adds to existing literature on culture's inherent multi-dimensionality, and seven new dimensions have been discussed. The multi-agent modelling and simulation of culture uses the seven-dimensional approach to understand how cultural belief-based equilibrium can emerge based on the relationships, communication, and influence idiosyncracies of individual agents in a complex organizational system.

The three initial simulation experiments show how different configurations of the same agent organization can result in different cultures, depending on when highly-influential agents-of-change are added to the system. Moreover, agent-oriented culture modelling has been demonstrated, and the results have shown how beliefs stabilize for a simple example, as a first step towards modelling more complicated cultures and diverse organizations.

Future work will target this direction and investigate how the addition or removal of groups of agents impacts culture, as in common organizational mergers and acquisitions, as well as testing different social-network configurations.

References

Alvesson, M.: Understanding organisational culture. Human Resource Management Journal **13** (2003)

Ashkanasy, N., Wilderom, C., Peterson, M.: Handbook of organizational culture & climate. Sage Publications, Inc (2000)

Bar-Yam, Y.: Dynamics of complex systems. Perseus Books Cambridge, MA, USA (1997)

Bicocchi, N., Ross, W., Ulieru, M.: A simulation modelling approach enabling joint emergency response operations. In: Systems Man and Cybernetics (SMC), 2010 IEEE International Conference on, IEEE (2010) 1832–1837

Clancey, W., Sachs, P., Sierhuis, M., Van Hoof, R.: Brahms: Simulating practice for work systems design. International Journal of Human Computer Studies **49** (1998) 831–866

Conte, R., Castelfranchi, C., Dignum, F.: Autonomous norm acceptance. In: Intelligent Agents V. Agent Theories, Architectures, and Languages: 5th International Workshop, ATAL'98, Paris, France, July 1998. Proceedings, Springer (2000) 66–66

Dignum, F., Dignum, V.: Emergence and enforcement of social behavior. 18th World IMACS / MODSIM Congress, Cairns, Australia (2009)

Dignum, F., Dignum, V., Jonker, C.: Towards agents for policy making. Multi-Agent-Based Simulation IX (2009) 141–153

Floridi, L.: What is the Philosophy of Information? Metaphilosophy **33** (2002) 123–145

Hatch, M., Cunliffe, A.: Organization theory: Modern, symbolic, and postmodern perspectives. Volume 379. Oxford University Press Oxford (1997)

Hofstede, G.: Culture's consequences: Comparing values, behaviors, institutions, and organizations across nations. Sage Publications, Inc (2001)

Hosseini, H.: A Reinforcement Learning Approach to Dynamic Norm Generation. Master's thesis, University of New Brunswick (2010)

Hosseini, H., Ulieru, M.: How Adaptable Are Your Norms? Leveraging Domain Knowledge to Learn Normative Behaviors. In Proceedings of Adaptive and Learning Agents (ALA) workshop at Tenth international joint conference on Autonomous Agents and Multi-Agent Systems (AAMAS). Taiwan; 2–6 May 2011. (2011)

Kroeber, A., Kluckhohn, C., Untereiner, W., Meyer, A.: Culture: A critical review of concepts and definitions. Vintage Books New York (1952)

Kroeber, A., Kluckhohn, C., Untereiner, W., Meyer, A., of American Archaeology, P.M., Ethnology: Culture: A critical review of concepts and definitions. Vintage Books New York (1963)

Morris, A., Ross, W., Ulieru, M.: Modelling Culture in Multi-agent Organizations. AMPLE Workshop Proceedings, Tenth international joint conference on Autonomous Agents and Multi-Agent Systems (AAMAS). Taiwan; 2–6 May 2011. (2011)

Rao, A., Georgeff, M.: BDI agents: From theory to practice. In: Proceedings of the first international conference on multi-agent systems (ICMAS-95), San Francisco, CA (1995) 312–319

Savarimuthu, B., Cranefield, S., Purvis, M., Purvis, M.: Role model based mechanism for norm emergence in artificial agent societies. In: Proceedings of the 2007 international conference on Coordination, organizations, institutions, and norms in agent systems III, Springer-Verlag (2007) 203–217

Ulieru, M., Verdon, J.: Organizational transformation in the digital economy. In: 7th IEEE International Conference on Industrial Informatics (INDIN), IEEE (2009) 17–24

Von Bertalanffy, L.: General system theory: Foundations, development, applications. G. Braziller New York (1968)

Chapter 3
Cross-Validation of Gaming Simulation and Multi-Agent Simulation

Gert Jan Hofstede, Catholijn M. Jonker, and Tim Verwaart

3.1 Introduction

The operation of social networks such as trade networks is a phenomenon of great complexity, since the social processes in such networks are composed of many sub-processes and variables at different levels of abstraction are involved: at the individual level there are personalities and skills; at the group level there are cultures, and possibly group labels. This chapter describes an approach to test theories about social processes and their differentiation across cultures. The approach is based on the combined application of gaming simulation and multi-agent simulation, where the gaming simulation is aimed to test predictions based on the theories, and the multi-agent simulation implements mechanisms according to the theories and thus is aimed to present a more detailed provisional explanation of the observed phenomena in the gaming simulation.

Gaming simulation is a technique that can be used to generate rich behavioral data by confronting subjects with realistic problem issues in complex social systems that provide ambiguous incentives (Duke and Geurts 2004; Klabbers 2008). Unlike, for instance, action research, gaming simulations abstract from reality and are set up to be repeatable and generalizable. Gaming simulations differ from other forms of experimentation in that they offer a social context in which group behavior and

G.J. Hofstede (✉)
Wageningen University, Postbus 9109, 6700 HB Wageningen, The Netherlands
e-mail: gertjan.hofstede@wur.nl

C.M. Jonker
Delft University of Technology, Mekelweg 4, 2628 CD Delft, The Netherlands
e-mail: c.m.jonker@tudelft.nl

T. Verwaart
LEI Wageningen UR, Postbus 29703, 2502 LS Den Haag, The Netherlands
e-mail: tim.verwaart@wur.nl

V. Dignum and F. Dignum (eds.), *Perspectives on Culture and Agent-based Simulations*, Studies in the Philosophy of Sociality 3, DOI 10.1007/978-3-319-01952-9_3,

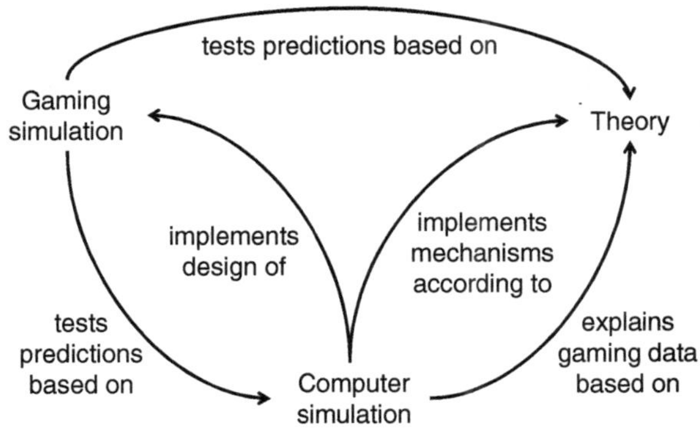

Fig. 3.1 Relationships between theory, gaming simulation, and computer simulation in a combined research approach

social networks can emerge. The question how realistic the gaming simulation results are, called the validation question, is not trivial. Gaming simulation professionals also notice the influence of group size, layout of physical space, time pressure, and other varied context factors.

Gaming simulation studies phenomena in a holistic way. Complementing it with computer simulation enriches the research by introducing an analytic approach. A gaming simulation is designed to test predictions based on a theory. According to Troitzsch (2004) a computer simulation is an application of a theory. It implements mechanisms according to the theory and thus does not only test the predictions, but also presents a provisional explanation of the data provided by the gaming simulation. Therefore, the validation of a simulation model, comparing the data it generates with the data observed in the gaming simulation, tests the theory about the mechanisms in a richer way than the observations do. Figure 3.1 summarizes the relationships between theory, gaming, and computer simulation.

Multi-agent simulation is a type of computer simulation that offers a natural paradigm for mechanism testing in combination with gaming simulations. A multi-agent model simulates the behaviors and interactions of individuals in a social system and the aggregate system behavior emerges from the individual's behaviors and interactions (Gilbert 2008). Developing a multi-agent simulation requires explicit formulation of the knowledge and hypotheses about the behaviors and interactions of agents, which are the mechanisms of the social system. Unlike equation-based simulations (for instance, based on system dynamics) multi-agent models are not based on assumptions about aggregate level system behavior. The aggregate behavior emerges from the individual-level behaviors and interactions, similar to the way it emerges in gaming simulations.

An agent-based model can be validated at aggregate and individual level. The process model of an individual agent operating in a realistic social context is necessarily more comprehensive than a model of a single theoretical concept. The process model of an agent is best described as a composition of sub-processes, each of which implement theories of different aspects of agent behavior, e.g. negotiation, trust, and deceit. The validation of an agent model may suffer from underdetermination: if a model explains the behavior of a system, it is not certain that it is the only model that can give an explanation (Sawyer et al. 1997). A main hypothesis of this chapter is that this uncertainty is reduced if not only the composition as an entity, but also sub-process models at lower compositional levels are validated.

Combined application of gaming simulation and multi-agent simulation proved a useful research method to develop descriptive models (i.e., implemented theories) of processes in trade networks (Meijer 2009). The gaming simulation and the multi-agent simulation are designed as counterparts: together they enable efficient data gathering to develop and formulate more detailed theories of human behavior in trade. The functions of the multi-agent simulation in the research method are:

1. To validate models of behavior induced from observations in gaming simulations;
2. To test hypotheses about dynamics of aggregated results in relation to parameter changes in individual behavior;
3. To select useful configurations for gaming simulations.

The first and second issues refer to what is above called mechanism testing; the third issue supports testing of theory-based predictions by gaming simulation.

The validation that is an essential part of this research method is complicated by the fact that social situations are culture-sensitive, i.e., the processes running underneath the social situation play out differently with players having different cultural backgrounds. Validation should prove that variations in system behavior in the gaming simulation between participants from different cultures correspond with variations in outcomes of the multi-agent simulation across those cultures.

This chapter introduces a cross-validation approach for a gaming simulation and its associated multi-agent simulation for culture-sensitive processes. As a leading case study, we describe the validation of an agent-based simulation of a gaming simulation, called the Trust And Tracing Game, in which negotiation and trade take place, including the possibility of deceit and of checking on it. The activities modeled in the simulated agents are partner selection, negotiation about a possible transaction, actual delivery after a deal has been closed, and maintaining beliefs about trade relations. The Trust And Tracing Game has been used for data collection in social research into the role of trust in supply networks (Meijer et al. 2008).

This chapter is organized as follows. Section 3.2 introduces the research project that serves as a case study for the work presented in this chapter – the Trust And Tracing Game – and the process model underlying the agent-based simulation. Section 3.3 discusses related work that combines gaming simulation or other experiments with human participants, with agent-based simulation. Section 3.4 discusses validation issues with respect to agent-based simulation and introduces elements of the cross-validation approach: composition of validated models, expert validation of cultural adaptation, sensitivity analysis, statistical validation at aggregated level, agent behavior, and micro gaming simulations. Section 3.5 introduces culture sensitive validation of processes in agents that are modeled to have a cultural background. Section 3.6 presents an example of an experiment in the on-going process of model validation. Section 3.7 proposes the concept of micro gaming simulation to validate aspects of the agent model in isolation. Section 3.8 discusses the role of sensitivity analysis in the validation and the importance of analyzing sensitivity at both individual and aggregated level. Section 3.9 concludes the chapter.

3.2 Background

The approach proposed in this chapter is illustrated by the example of the Trust And Tracing game. This gaming simulation is a research tool for supply chain and network studies. It places the choice between relying on trust versus spending money on complete information in trade environments at the core of a social simulation game. The game is used both as a data gathering tool about the role of reputation and trust in various types of business networks, and as a tool to make participants reflect on their own daily experiences in their respective jobs (Meijer et al. 2006).

A brief description of the Trust And Tracing game will be given here. An extensive description is available in (Meijer et al. 2006). In the game 12 up to 25 participants play roles in a supply network. There are roles of producers, middlemen, retailers and consumers (see Fig. 3.2). Sellers of a commodity have more information about the quality of the goods than buyers, since quality is invisible and only known by the producers. This leads to information asymmetry and the opportunity for deceit.

Each player receives (artificial) money. The producers receive an initial amount of goods. The good traded is a sealed envelope that comes in 3 different types (colours) and each of the types in two qualities (high and low). A ticket covered in the envelope (so it is not visible) represents quality. Producers are informed which envelopes are high quality and which are low. The only person in the game allowed to open an envelope is the tracing agency. The game leader acts as a tracing agency and can, on request and at the cost of a fee, determine product quality.

The goal of producers and traders is to maximise profit. The consumers' goal is to maximise satisfaction. Table 3.1 specifies satisfaction values of each good for a consumer (utility points).

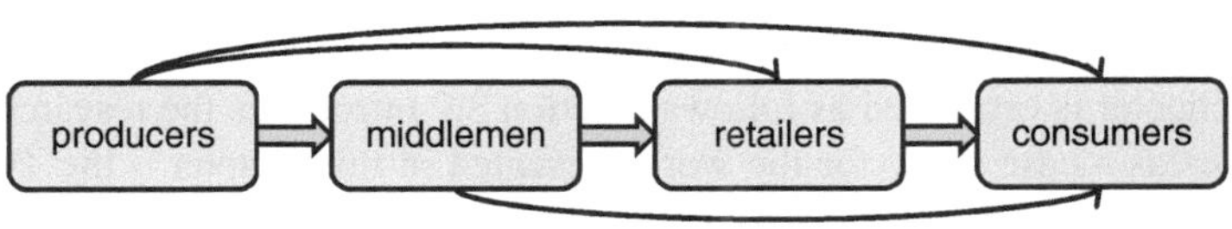

Fig. 3.2 Supply network configuration with commodity flows in the Trust And Tracing Game

Table 3.1 Consumer satisfaction value by type and quality of the commodity

	Commodity type		
Quality	Blue	Red	Yellow
Low	1	2	3
High	2	6	12

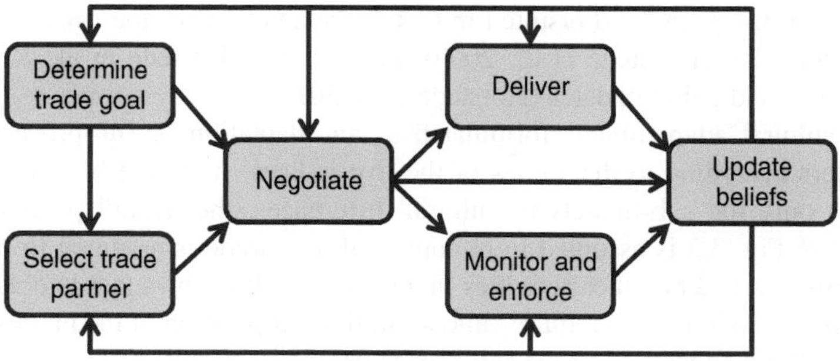

Fig. 3.3 Sub-processes and internal information flow of trading agents

An agent buying a high quality envelope that has not been traced yet takes a risk, as he cannot know the real quality. The buyer can check afterwards by doing a trace at the tracing agency, but this costs money. Tracing is cheaper early on in the network than for consumers. When consumers prefer traced goods (certified high quality) it would be economical to let a middleman do the trace and sell the traced product throughout the network along with the certificate.

Successful deceit is beneficial for a seller since it results in additional income. However, if the deceit is discovered the cheater has to pay a fine. Resellers of cheated products who did not check the quality themselves but acted in good faith have to pay a smaller "ignorance" fine.

The gaming simulations played with groups of participants from different cultural backgrounds produced different outcomes; for instance, patterns of deceit and trust differed considerably between Dutch and American business school students (Meijer et al. 2006). This implies that an agent-based simulation of the game must take culture into account.

The research with the Trust And Tracing Game combines gaming simulation and multi-agent simulation. The process model for trading agents acting in the multi-agent simulation is given in Fig. 3.3. It is based on the sources of transaction cost according to transaction cost economics (Williamson 1998): searching, bargaining, and contract drafting in the pre-contract phase of transactions and, taking the risk of partners' opportunistic behavior with respect to delivery into account, monitoring and enforcing in the post-contract phase.

The plans that the agents execute for fulfillment of the sub-processes in the agent-based simulation, are based on validated models taken from literature on social sciences and artificial intelligence. A problem with these models is that they have been validated locally in one particular culture, while the behavior in the game has been shown to be different across cultures. So, realistic simulation requires culturally differentiated agents.

Cultures can be characterized by multi-dimensional indices. The most widely used are Geert Hofstede's five dimensions of national cultures (Hofstede 2001). A method based on an expert systems approach was developed that adapts the sub-process models using the values of the five dimension indices (Hofstede et al. 2010).

Application of this method resulted in cultural adaptations of the agent models of partner selection (Hofstede et al. 2009), negotiation (Hofstede et al. 2012), and deceit, trust, and belief updates (Hofstede et al. 2011).

The cultural adaptation is formulated as an adaptation of sub-process model parameters according to the values of the five indices. It should be noted that by adapting only the sub-models to cultural differences, the overall process model depicted in Fig. 3.3 is assumed to be universal, i.e., trade is assumed to comprise these activities and no other activities in any culture. It follows that two aspects of the multi-agent simulation require validation: the composition of the process model and the cultural adaptation of the sub-processes.

The research with the Trust And Tracing Game combines gaming and multi-agent simulation. The multi-agent simulation is instrumental to the gaming simulation. The approach can be formulated as follows. The theory T formed on the basis of the literature and the gaming simulation G, has to be formalized in such a way that an implementation A can be constructed that underlies the multi-agent simulation. This formalization step requires enriching the theory to the level of the individual. The multi-agent simulation A must be validated to sufficiently reproduce outcomes of gaming simulation G, but at the same time it acts as a test for the theory T, in the sense that mechanisms implemented in A should provide similar results as the gaming simulation G. Furthermore, the multi-agent simulation A can be used to explore the space of possible experimental setups, which enables the identification of those setups that would most critically test the theory T when played in the gaming simulation G. These setups can be selected on the basis of a sensitivity analysis that identifies the circumstances under which the behavior of A shows strong changes. These are the experimental setups for which gaming simulations can best be played for testing hypotheses based on T. The data and conclusions of these new gaming simulations might comply with the standing theory or might lead to revisions. Theory revisions lead to adaptations of the multi-agent simulation (or the game design) and form the beginning of a new cycle.

A validation method that supports the combined method of gaming- and multi-agent simulation should address the validity of the application of theory T in the gaming construct G, and in the agent-based implementation A. Note that the validation of G and A are instrumental, but not enough to validate T as a theory of the studied real-life social situations. As the multi-agent simulation A is designed in interaction with the results of G and the theory T, there is a need for cross-validation of G and A.

3.3 Related Work

The idea of combining gaming with agent-based simulation goes back to the work of Barreteau et al., who applied this combined approach to natural resource management in developing countries (see, e.g., Barreteau et al. 2001). Barreteau et al. discern the combined application of gaming and agent-based simulation for training purposes, for research purposes, and for decision support processes. For research purposes, they claim that agent-based simulation leads to the production and validation of theories, where validation relies upon finding a match between observed and simulated results (replicative and predictive validity) as well as between modelled and real processes (structural validity).

Barreteau et al. (2001) address two problematic issues with respect to validation. First, results to be validated are difficult to observe in the real process. Second, the capacity of the set of selected processes to display the same patterns of interactions must be tested. According to the authors, gaming appears, a priori, to offer an attractive solution. It is possible to record and partly control its progress in order to compare it with multi-agent simulations and if players and/or observers are stakeholders in the real process, they provide a means to reflect on their behaviour in it in reality. The stakeholders' discussions contribute to the model validation in their companion modelling approach (Barreteau et al. 2003).

To a great extent, our work is similar to that of Barreteau et al. However, it differs significantly in two respects. First, our work focuses on the influence of culture. The participants in a companion modelling activity can reflect on their behaviour in the processes displayed in Fig. 3.3, but they are not cultural experts who can reflect on the influence of their cultural background on their behaviour. We need expertise on culture for structural validation and face validation of results. Second, given that the social context is most relevant when doing cross-cultural research, in our approach the agent-based simulation is instrumental to the gaming simulation: it aims to simulate the game (the research tool), not the real world. Barreteau et al. aim to model the real world in the agent-based simulation and use the gaming simulation for validation of the agent-based model.

Guyot et al. (2007) and Heckbert (2009) both describe a method of experimental data collection where human subjects play a role in a multi-agent simulation. In these experiments, the gaming simulation environment and other players are computer-simulated. The experiments focus on the validation of particular aspects of the agent model. For research into trade networks where aspects like interpersonal trust and culture are relevant, the rich social context of human interaction is an essential element of the gaming simulation. It cannot be replaced by a computer simulation. However, for validation of the cultural adaptation of the sub-process models, experiments with human participants in an agent environment can be a valuable extension to the data collection toolbox (Hofstede et al. 2008).

3.4 Cross-Validation

Boero and Squazzoni discuss empirical calibration and validation of three types of agent-based models, studying two examples from literature of calibration and validation for each of the three model types (Boero and Squazzoni 2005). On one end of the continuum they identify *case-based* models that are specific for empirically circumscribed systems. For these models the micro-level data gathering strategy is the main issue. On the other end of the continuum are *theoretical abstractions*, which refer to general social mechanisms. Those models illustrate a theoretical concept, for which data gathering and validation are an on-going process. In between the extremes is a third class of models: *typifications*, which intend to investigate theoretical properties of a particular class of phenomena. The model of the Trust and Tracing Game classifies as a typification.

According to Boero and Squazzoni, typifications can be used to embed a case-based model into a wider theoretical reference, or vice-versa: a representative case can be used as an empirical test for a typification (Boero and Squazzoni 2005). The latter is exactly what the role of gaming simulations is in the simulation of the Trust And Tracing Game: data gathering in representative cases to test a theory and the agent model which applies that theory.

Gilbert refers to typifications as *middle range models* that "*aim to describe the characteristics of a particular social phenomenon, but in a sufficiently general way that their conclusions can be applied widely*" (Gilbert 2008). He explains that by the generic nature of these simulations their behaviour cannot be exactly compared with any observable instance, but that one must be satisfied with qualitative resemblance. For instance, for our simulation we would be satisfied with resemblance of *stylized facts* of the form "the tracing frequency in games with American business school students is higher than it is in games with Dutch business school students".

Resemblance of stylized facts at aggregate level is not a sufficient validation. Takadama et al. describe the validation of an agent model of the sequential bargaining game by means of experiments with human participants (Takadama et al. 2008). They compare the results of Q-learning agent types with different action selection mechanisms. At the macro-level, output distributions of all action selection types converge to distributions found in human experiments. However, only the learning curve of one of the agent types (the type using a Boltzmann distribution for action selection) resembles the learning curve of human subjects.

The work of Takadama et al. stresses the relevance of process validation in multi-agent simulations, in order to reduce the under-determination problem. Janssen and Ostrom (2006) also argue that given empirical problems with data collection and the explicit inclusion of cognitive, institutional and social processes in agent-based modelling, statistical performance is not sufficient for validation. Other criteria for validation of agent-based models they mention are:

- Plausibility of the model, given the understanding of processes,
- Understanding why a model performs well,
- Better understanding of empirical observations gained through the model,
- Stakeholder validation of model behaviour.

It should be noted that without the results of a thorough sensitivity analysis the issues raised by Janssen en Ostrom cannot be covered. According to Gilbert, sensitivity analysis to acquire an understanding of model behaviour is an essential element of the validation of agent-based models (Gilbert 2008).

Moss and Edmonds used the term cross-validation for an agent-based model validation approach where the model is based on data and assessments provided by domain experts and stakeholders, and cross-validated against statistical data at the macro level and qualitative research results at the micro level (Moss and Edmonds 2005).

According to Moss and Edmonds, agent-based simulation differs fundamentally from statistical research and qualitative approaches in that it allows for validation of the representation of agent behaviour as well as characteristics of aggregate time series (Moss and Edmonds 2005). As a basis for validation Moss (2002) refers to data collection at individual level (interviews and surveys) to collect data about distributions of actions, their relations with demographics, and patterns of social interaction; an agent based simulation can be validated at any level of detail. Further, that author refers to the application of validated concepts from social psychology and cognitive science and stakeholder participation as elements of model validation.

We take a similar cross-validation approach as Moss and Edmonds (2005), with the following elements:

- Apply validated results from social science to model sub-processes and culture; thus, only the top-level process model (i.e. the composition of the sub-processes) and the integration of culture in the sub-processes remain to be validated;
- Perform expert validation of the integration of the model of culture with the sub-process models;
- Perform sensitivity analysis to discover model behaviour under different parameter settings and compare the results with theory-based expectations (face validity);
- Compare gaming and agent-based simulation statistics, such as the number of successful transactions, average quality level of traded commodities, frequency of failed negotiations;
- Compare agent behavioural characteristics, such as loyalty, negotiation results, deceit, explicit distrust;
- Design micro-level games to test the integration of culture in sub-process models, possibly applying human participation in multi-agent simulations.

It is important to note that this approach does not comprise a global calibration of the agent-based model to gaming results. First, like for most agent-based simulations, the high number of parameters would inevitably lead to over-fitting. We must rely on calibration using other data at the sub-process level. The results of the sensitivity analysis can be helpful in selecting actual parameter values, but these must be in a plausible range according to the social sciences results that are at the basis of the sub-process models. Second, as Gilbert has pointed out, for a middle range model one must be satisfied with qualitative resemblance (Gilbert 2008). A middle range model cannot be expected to make accurate forecasts.

3.5 Culture-Sensitive Validation of a Process

The cross-validation method introduced in this chapter is an iterated approach, in which each iteration consists of three steps: first, validate the model irrespective of culture (i.e., in some cultural setting), then for a number of culturally homogeneous populations, and subsequently vary culture by comparing gaming simulations played with participants with different cultural background. The iteration is done for every compositional level of the process model. In this chapter we focus on the second step of each iteration, in which culture plays are role.

In the multi-agent simulation, culture is modeled according Hofstede's model of culture. Hofstede classifies national cultures according to five indices on the dimensions presented in Table 3.2.

National cultures can be compared on these dimensions. Figure 3.4 displays the scores the Netherlands and the USA. The scores are rather similar, except for the dimension of masculinity versus femininity. If results from gaming and multi-agent simulations match for players with a cultural background in these countries, the confidence in the model to correctly represent the effects of that dimension,

Table 3.2 Hofstede's five dimensions of culture (Hofstede 2001)

Dimension	Definition
Power distance	"The extent to which the less powerful members of institutions and organizations within a country expect and accept that power is distributed unequally" (Hofstede 2001), p. 98
Uncertainty avoidance	"The extent to which the members of a culture feel threatened by uncertain or unknown situations" (Hofstede 2001), p. 161
Individualism and collectivism	"Individualism stands for a society in which the ties between individuals are loose: Everyone is expected to look after him/herself and her/his immediate family only. Collectivism stands for a society in which people from birth onward are integrated into strong, cohesive in-groups, which throughout people's lifetime continue to protect them in exchange for unquestioning loyalty" (Hofstede 2001), p. 255
Masculinity and femininity	"Masculinity stands for a society in which social gender roles are clearly distinct: Men are assumed to be assertive, tough, and focused on material success; women are supposed to be more modest, tender and concerned with the quality of life. Femininity stands for a society in which gender roles overlap: Both men and women are supposed to be modest, tender and concerned with the quality of life." (Hofstede 2001), p. 297
Long- versus short-term orientation	"Long term orientation stands for the fostering of virtues oriented towards future rewards, in particular, perseverance and thrift. Its opposite pole, short term orientation, stands for the fostering of virtues related to the past and the present, in particular, respect for tradition, preservation of 'face' and fulfilling social obligations" (Hofstede 2001), p. 359

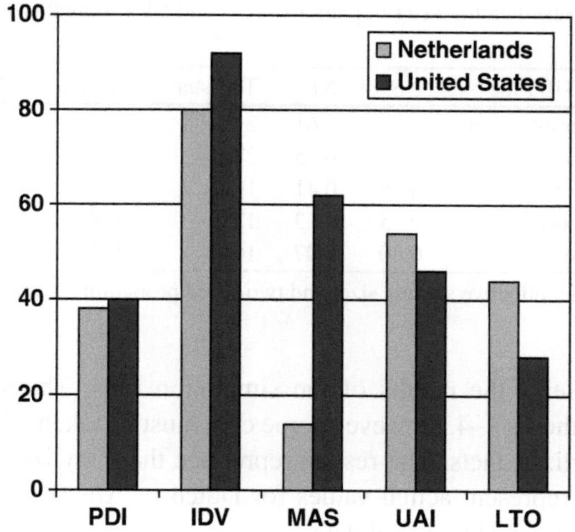

Fig. 3.4 The cultural indices of The Netherlands and the USA (Data source: (Hofstede 2001))

grows. Validation of middle range models is an on-going process. By successively conducting this kind of experiments for different cultural configurations, the validation level is increased.

3.6 Example of Experimental Results

This section discusses an example on the basis of experiments with the Trust And Tracing Game, reported by Meijer et al. (2006). In gaming simulations with the Trust And Tracing game with business school students in The Netherlands and in the USA, the American students showed more eager to win, traded higher quality, seized opportunities to cheat, and expected their opponents to do so too, so they traced more frequently. Also in the USA a greater fraction of high quality transactions was certified, i.e. traced up-front by the supplier. The reason for this is that the tracing fee for suppliers was lower than it was for customers. The players discovered that with a high tracing probability, it was efficient to have the suppliers trace in advance and use the tracing report as a quality certificate. The following hypotheses can be formulated to be tested against simulation results:

1. The quality ratio (top quality transactions/all transactions) is higher in USA than in the Netherlands
2. The certification ratio (certified transactions/top quality transactions) is higher in USA than in the Netherlands
3. The defection ratio (untruthful deliveries/uncertified top quality transactions) is higher in USA than in the Netherlands
4. The tracing ratio (traces/uncertified top quality transactions) is higher in USA than in the Netherlands

Table 3.3 Test data for 310 run pairs for USA and NL (Source: (Hofstede et al. 2011))

Average of 310 runs	USA	NL	Test stat.[a]	Sample[a]	Probability[a]
Number of transactions	72	61	219	302	<0.001
Quality ratio	0.37	0.15	277	285	<0.001
Certification ratio	0.48	0.41	191	281	<0.001
Defection ratio	0.25	0.13	128	154	<0.001
Tracing ratio	0.40	0.07	169	177	<0.001

[a]Test statistic, effective sample size, and two-sided probability level for sign test

Table 3.3 presents the results of the simulation runs. The simulation results confirm the hypotheses 1–4. However, some care must be taken. The hypotheses are formulated as stylized facts. The results reproduce these stylized facts, but cannot be interpreted to represent actual values for Dutch or American behaviors, as the model has not been tuned to actual data.

As discussed in Sect. 3.3, these results increase the confidence that the model correctly represents the influence of the masculinity versus femininity dimension.

3.7 The Next Compositional Level: Micro Gaming Simulations

When playing the Trust And Tracing Game in different cultural settings, it was observed that culture matters for the outcomes of the game (Meijer et al. 2006). What is the cause of these differences? The process model presented in Fig. 3.3 is universal. In order to trade, people need to find a partner, come to an agreement about the transactions, have the opportunity to defect and will consequently have to be prepared to monitor and enforce the delivery according to the contract. These activities are the essence of trade. The difference is in the way these activities are performed in different cultures.

As it was formulated earlier in this paper, a sufficiently formalized implementation A of a theory T – formed on the basis of the literature – and the gaming simulation G underlies the multi-agent simulation. Validation of G, A, and their interaction is no trivial matter, as the agent-based model underlying the implementation A combines a variety of models of theories that concern some aspect of the social situation, e.g., in trade we have to consider partner selection, negotiation, and delivery. Each of these sub-models has to be validated in separation as much as possible. Furthermore, the validated sub-models have to be validated in combination. Composing sub-models that have not yet been validated, and subsequently attempting to validate the composition would lead to underdetermination. In the preceding sections the validation of the composed model is considered. The present section considers the validation of the components, in particular their adaptation to culture.

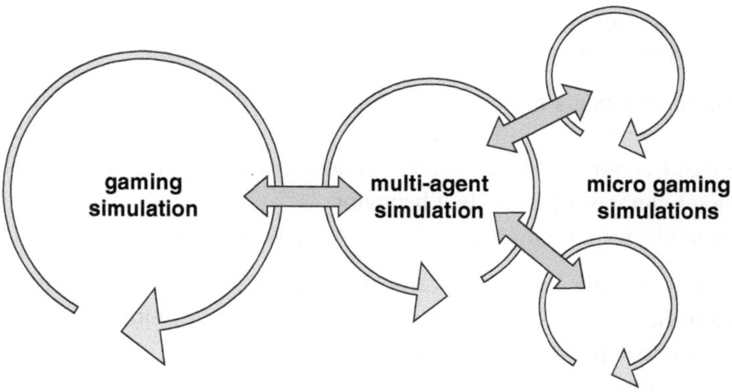

Fig. 3.5 Validation relations between gaming simulation, agent-based simulation, and micro gaming simulations

Typically the sub-models in question have been developed in mono-cultural settings, whereas a social activity such as international trade is per definition a multi-cultural setting. This illustrates that validated sub models with respect to multi-cultural settings are not to be found. Of course, the problem of under-determination holds in general. To reduce this problem (in this example with respect to culture) the sub-models can be subjected to an additional validation process (in this example that would be testing it with humans of different cultures). We conclude that any compositional modeling approach for a complex system S runs into the problem of combining sub-models for subsystems that are underdetermined with respect to the modeling challenge of the whole system S.

In the agent model, the cultural differentiation is incorporated in the models of the sub-processes. The models applied in the sub-processes have been validated in experiments with human participants (this was a selection criterion for these models, a priori), but the adaption for culture has not. Cross-validation of statistics and behaviour in gaming versus multi-agent simulation increases our confidence in the correctness of the agent model and the integration of culture in the sub-process models. However, the risk of under-determination remains. That risk would be reduced if the sub-models could be tested in isolation. That is the purpose of the micro gaming simulations we propose in this section.

An additional feature of such a micro gaming simulation is that it gives a truth test of elements of the agent-based simulation with real people. Thus the system is tested at multiple levels: (1) the system behavior of the gaming simulation emerging from micro behaviors, (2) behaviors of individuals in the gaming simulation, and (3) elements of the individual's behaviors in isolation (see Fig. 3.5).

It is to be kept in mind that the long-term aim of the research for the case study is to develop a theory of human behavior in trading situations. The gaming simulation and the agent-based simulation are designed as counterparts: together they enable us to efficiently and effectively gather research data to develop and formulate more detailed theories of human behavior. The functions of the agent-based simulation in the research approach are:

1. Validation of models of behavior induced from observations in gaming simulations
2. Testing of hypotheses about system dynamics of aggregated results in relation to parameter changes in individual behavior
3. Selection of useful configurations for gaming simulations

The micro gaming simulations are designed to validate the decision models of the agent-based simulation. There is little experience with such an approach. A few conceptual papers have been published (e.g. Guyot et al. 2007; Heckbert 2009), but no discipline has evolved. The context of the micro gaming is not as rich as that of the full gaming simulation, but since the essential element to be validated is the cultural adaptation of the sub-process models, some social context must be offered. Hofstede et al. have proposed a micro gaming simulation to test aspects of negotiation about the procurement of a second hand car (the Lemon Car game) (Hofstede et al. 2008). This game is currently being further developed for application in secondary education. No results can now be reported on this work in progress. The concept of the game allows for data collection in live situations and through the internet.

3.8 Sensitivity Analysis in Agent-Based Model Validation

The role of sensitivity analysis in agent-based simulation is to evaluate the fit between the theory and its implementation in the agent-based model (Gilbert 2008). In many types of simulation, sensitivity analysis is seen as a technical activity, aiming to improve the correctness of model implementation and the accuracy of forecasts made with the model. In the case of agent-based models, the relation between model formulation and output is hard to predict and sensitivity analysis also includes the discovery of emerging properties at the macro level. Therefore, sensitivity analysis must be performed at the agent level and at the macro level.

Burgers et al. (2010) have shown that, for the Trust And Tracing Game, sensitivity at the macro level and at the individual level can differ considerably. For instance, some parameter change may have little effect on the total number of transactions in the game, even if it greatly affects the performance of individual agents in different ways; other agents may step into the place of agents that do not perform well under the changed parameter settings. Sensitivity analysis at the individual level is complex. The behaviour of an agent does not depend solely on its own parameter settings. It is also influenced by the behaviours of the agents it interacts with. Burgers et al. conclude that methods for sensitivity analysis at the individual level have to be developed.

The multitude of parameters that are involved complicates the sensitivity analysis of agent-based simulations. In fact, every agent participating in the simulation has its individual parameter set. So, one has to work in a stylized way, using one or several groups of homogeneous agents, or generating the parameters from a distribution of which the parameters are taken as the input parameters to the simulation.

Sensitivity analysis at macro level can be performed by well-developed methods (Saltelli et al. 2000). However, in the case of culture-sensitive agents the analysis is far from straightforward – and time-consuming for the researcher. Strong interactions prevail between parameters of culture and other parameters. Burgers et al. (2010) conclude that straightforward sensitivity analysis of the Trust And Tracing multi-agent model based on the method of Jansen et al. (1994) can only be performed in a fixed cultural setting. Thus, ideally, sensitivity analysis should be performed for every new cultural setting the model is applied in.

Although it is a laborious undertaking, sensitivity analysis pays off. It contributes to the validation criteria mentioned by Janssen and Ostrom (2006):

- It provides face validity of results, given the understanding of processes,
- Sensitivity analysis contributes to understanding why a model performs well,
- Results provide better understanding of observations gained through the model,
- The results of sensitivity analysis can be input to stakeholder validation.

Moreover, sensitivity analysis identifies regions in parameter space where the model is particularly sensitive or insensitive to parameter changes. This information can be used in the experimental set-up of gaming simulations.

3.9 Discussion and Conclusions

From trade process in the world to a gaming simulation is an abstraction that enables the researcher to execute repeatable experiments, i.e. a controlled setting with the essentials of trading in place, maintaining a social context in the experiments. As the required number of experiments with the gaming simulation, and the overhead of performing an experiment are prohibitive, an agent-based simulation is developed. However, this way another major abstraction is performed. This chapter shows a method with which it can be determined whether all important aspects of the gaming simulation are covered in a way that the behaviors seen in the gaming simulation also occur in the agent-based simulation and in a comparable way. Furthermore, our method determines whether the important sub-models of the agent-based model are valid with respect to real human behavior. This method is important as the agent-based simulation is intended to function as a way to:

1. Validate models of behavior induced from observations in gaming simulations;
2. Test hypotheses about dynamics of aggregated results in relation to parameter changes in individual behavior;
3. Select useful configurations for gaming simulations by performing sensitivity analysis on the agent-based analysis.

The approach we introduced for these purposes consists of:

- Compositional agent modelling, using components based on validated results from social science, and a validated theory of culture, so that only the selection of components and their integration need to be validated;
- Expert validation of the integration of the cultural theory into the components;
- Sensitivity analysis to discover model behaviour under different parameter settings and compare the results with theory-based expectations (face validity);
- Validation of agent-based simulation statistics against gaming statistics on aggregate level;
- Validation of agents' behavioural characteristics with those observed in the gaming simulation;
- Micro gaming simulations to test the integration of culture in sub-process models, possibly applying human participation in multi-agent simulations.

The micro gaming simulations have less context than the full simulation and focus on one or some aspects of one or some sub-processes. Thus, they contribute to structural validation of the mechanisms and reduce the under-determination of the full gaming simulation. Future research should develop micro gaming simulations for validation of aspects of the cultural adaptability of the Trust And Tracing agent model. For this purpose, computer-supported gaming simulations – like the one experimented with in the Lemon Car game (Hofstede et al. 2008) – can be used. Guyot has called this approach participatory multi-agent simulation (Guyot et al. 2007).

For a middle-range model like the culture-sensitive Trust And Trading game simulation, validation is an on-going process. It is infeasible to validate such a model for all possible configurations, but confidence in this type of models grows as they are validated in more and diverse situations. Future research should validate the model for other configurations than discussed in this paper.

References

O. Barreteau, F. Bousquet, J. Attonaty. Role-playing games for opening the black box of multi-agent systems: method and lessons of its application to Senegal River Valley irrigated systems. Journal of Artificial Societies and Social Simulation 4 (2), 5 (2001)

O. Barreteau et al. Our companion modeling approach. Journal of Artificial Societies and Social Simulation 6 (1), 1 (2003)

R. Boero, F. Squazzoni. Does Empirical Embeddedness Matter? Methodological Issues on Agent-Based Models for Analytical Social Science. Journal of Artificial Societies and Social Simulation 8 (4), 6 (2005)

S.L.G.E. Burgers, G.J. Hofstede, C.M. Jonker, T. Verwaart. Sensitivity Analysis of an Agent-Based Model of Culture's Consequences for Trade. In: M. Li Calzi et al. (eds.) Progress in Artificial Economics, Lecture Notes in Economics and Mathematical Systems 645, pp. 253–264. Springer (2010)

R.D. Duke, J.L.A. Geurts. Policy Games for Strategic Management. Dutch University Press (2004)

N. Gilbert. Agent-based models. SAGE Publications (2008)

P. Guyot, A. Drogoul, S. Honiden. Multi-Agent Participatory Simulations Between Experimental economics and Role-Plaing Games. In: T. Terano et al. (eds) Agent-Based Approaches in Economic and Social Complex Systems IV. Springer (2007)
S. Heckbert. Experimental economics and agent-based models. 18th World IMACS/MODSIM Congress, pp. 2997–3003 (2009)
G. Hofstede. Culture's Consequences, Second Edition. SAGE Publications (2001)
G.J. Hofstede, T. Verwaart, C.M. Jonker. Lemon car game. In: E. Bagdonas et al. (eds.) GAMES: virtual worlds and reality; Selected papers of ISAGA 2008. Kaunas University of Technology (2009)
G.J. Hofstede, C.M. Jonker, T. Verwaart. Simulation of Effects of Culture on Trade Partner Selection. In: C. Hernández et al. (eds) Artificial Economics: The Generative Method in Economics, pp. 257–268. Springer (2009)
G.J. Hofstede, C.M. Jonker, T. Verwaart. Cultural Differentiation of Negotiating Agents. Group Decision and Negotiation 21 (1), 79–98 (2012)
G.J. Hofstede, C.M. Jonker, T. Verwaart. A Cross-cultural Multi-agent Model of Opportunism in Trade. In: N.T. Nguyen, R. Kowalczyk (eds.) Transactions on Computational Collective Intelligence II, LNCS 6450, pp. 24–45. Springer (2010)
G.J. Hofstede, C.M. Jonker, T. Verwaart. Computational modeling of culture's consequences. In: T. Bosse et al. (eds.): MABS 2010, LNAI 6532, pp. 136–151. Springer (2011)
M.J.W. Jansen, W.A.H. Rossing, R.A. Damen. Monte Carlo estimation of uncertainty contributions from several independent multivariate sources. In: J. Grasman, G. van Straten (eds.) Predictability and Nonlinear Modelling in Natural Sciences and Economics. Kluwer (1994)
M.A. Janssen, E. Ostrom. Empirically based, agent-based models. Ecology and Society 11 (2), 37 (2006)
J.H.G. Klabbers. The magic circle: principles of gaming and simulation 2nded. Sense (2008)
S.A. Meijer, G.J. Hofstede, G. Beers, S.W.F. Omta. Trust And Tracing game: learning about transactions and embeddedness in the trade network. Journal of Production Planning and Control 17 (6), 569–583 (2006)
S.A. Meijer, G.J. Hofstede, G. Beers, S.W.F. Omta. The organization of transactions: research with the Trust And Tracing Game. Journal on Chain and Network Science. 8 (1), 1–20 (2008)
S. Meijer. The organization of transactions: Studying supply networks using gaming simulation. Ph.D. thesis. Wageningen Academic Publishers (2009)
S. Moss. Agent based modelling for integrated assessment. Integrated Assessment 3 (1), 63–77 (2002)
S. Moss, B. Edmonds. Sociology and simulation: Statistical and qualitative cross-validation. American Journal of Sociology 110 (4), 1095–1131 (2005)
A. Saltelli, K. Chan, E.M. Scott. Sensitivity Analysis. Wiley (2000)
K.R. Sawyer, C. Beed, H. Sankey. Underdetermination in Economics. The Duhem-Quine Thesis, *Economics and Philosophy 13*, 1–23 (1997)
K. Takadama, T. Kawai, Y. Koyama. Micro- and Macro-Level Validation in Agent-Based Simulation: Reproduction of Human-Like Behaviors and Thinking in a Sequential Bargaining Game. Journal of Artificial Societies and Social Simulation 11 (2), 9 (2008)
K.G. Troitzsch. Validating simulation models. In: G. Horton (ed) Proceedings of the 18th European Simulation Multiconference. SCS, Erlangen (2004)
O.E. Williamson. Transaction cost economics: how it works; where it is headed. De Economist 146 (1), 23–58 (1998)

Chapter 4
Modelling Culture Through Social Activities

Rubén Fuentes-Fernández, Jorge J. Gómez-Sanz, and Juan Pavón

4.1 Introduction

Computer simulations facilitate the study of culture phenomena through the formal specification of models of the behaviour of societies and the analysis of their evolution. Despite its advantages, the use of computer simulation presents nowadays some methodological and technical limitations.

A main issue is that most of the concepts that appear in Social Sciences do not have an intuitive translation to computational primitives. When developing simulations, complex correspondences between social and computational concepts must be defined based on knowledge, interpretations and hypotheses about their meaning and behaviour. Moreover, culture is a phenomenon where different overlapped contexts affect potentially large populations in ways that are frequently non-quantifiable and indirect. Models must be able to represent these side effects in a plausible way with influence in the simulation while considering scalability issues.

To carry out these tasks, social scientists need some expertise on the computational formalisms that are used to specify simulation models, like mathematical, logical and programming languages. The common way to address this issue is involving experts in computer simulation (Drogoul et al. 2003). However, this solution raises concerns about potential misunderstandings among participants due to their different backgrounds. This situation derives in difficulties to model and validate simulations, and to guarantee that they and their results correspond to the initial requirements.

R. Fuentes-Fernández (✉) • J.J. Gómez-Sanz • J. Pavón
GRASIA, Universidad Complutense de Madrid, Madrid, Spain
e-mail: ruben@fdi.ucm.es; jjgomez@fdi.ucm.es; jpavon@fdi.ucm.es

V. Dignum and F. Dignum (eds.), *Perspectives on Culture and Agent-based Simulations*, Studies in the Philosophy of Sociality 3, DOI 10.1007/978-3-319-01952-9_4,

A way to mitigate these problems would be trying to acquire the concepts that social scientists use and implement them with software, in an incremental and documented manner that can be adapted to different simulation problems (Edmonds 2003). The result would be simulation frameworks where social scientists could be able to express their culture models with social concepts. The computer researchers in charge of developing the software infrastructure would need to gain certain understanding about social research, but they could validate their work through multiple simulations. Current approaches based on simulation platforms, such as Netlogo, MASON and Repast (Railsback et al. 2006), do not meet these requirements. They enforce specific modelling approaches, are difficult to adapt to other premises, and are focused on programming.

Our research pursues developing a framework that would facilitate analysis about societies and cultures by providing a way to work with models made of social theory artefacts. The framework bases on the Activity Theory (AT) (Leontiev 1978), whose approach to analyse human behaviour focuses on how people act in their physical and socio-cultural context, and the development of these acts. The socio-cultural context covers the artefacts and meanings used by a group of people in their interactions. Interactions are conceptualised as networks of individual activities connected through the mental and physical artefacts they use and modify. These interactions and their contexts constitute the *social activities*. The framework enables progressively defining the elements of the social activities, which turns out to define the culture of their society.

The framework includes a graphical modelling language for social activities, rules that determine the evolution of its specifications over time, and techniques for the analysis of properties in them. In order to facilitate the development of specifications and reusing knowledge, it encourages the definition of *social properties* as specification blocks with all the relevant information to define a given aspect of the system. The use of AT related abstractions aims to provide social scientists with a modelling language close to their fields; the availability of a modelling language with an execution model facilitates computer researchers the refinement of the abstract specifications to simulations, and the extension of the framework if required.

This chapter illustrates the previous elements and their use through a case study about the impact of group culture in the way people deal with couple problems (Swidler 2003). The next section motivates the adoption of AT as a comprehensive framework to study culture. Section 4.3 presents the key aspects of AT. Section 4.4 describes the modelling framework. Section 4.5 develops the specification of the case study and shows the use of its models to study the effects of cultural contexts in individual behaviours. The chapter concludes with an analysis of the current strengths and limitations of the framework, as well as a description of ongoing and future work.

4.2 Culture in Social Sciences and Activity Theory

The Activity Theory (AT) (Leontiev 1978; Vygotsky 1978) is a comprehensive paradigm from Social Sciences. It supports a holistic approach to study human phenomena that integrates aspects ranging from the individual processes of cognition and perception to social issues. For this reason, researchers find in AT a powerful tool to study culture (Engeström 1987; Vygotsky 1978).

Culture is a controversial object of study in Social Sciences (Boyd 1988). There is agreement on its conceptualization as transmitted heritage in a given human society (D'Andrade 1995), but not on what constitutes it or the mechanisms for its development and transmission. For instance, some schools focus on the ideas that people share (Bratus et al. 1983) and others on observable behaviours and artefacts (D'Andrade 1995). From an AT perspective, culture is a system of shared artefacts socially built and communicated by a group. The notion of artefact is broad, as it covers both physical objects (e.g. machines or artistic creations) and mental constructions (e.g. language or norms), and their origin can be either in people activities or in the environment. This approach is very general, and suitable to accommodate most of paradigms as long as researchers model focusing on activities.

There are no global and centralized activities of a group in charge of building culture, as this happens as a consequence of the individuals' activities (Shore 1996). These activities are not necessarily coordinated: they work on certain artefacts and cause indirect changes on others, which in turn trigger new changes. Group activities constitute a convenient means to abstract certain individuals' activities in groups. AT supports the analysis of both individual and group activities with the hierarchical decomposition of activities at different levels of abstraction.

All these activities are mostly determined by the group culture. In this sense, the culture becomes the container and product of the socio-historical development and characteristics of its group (D'Andrade 1995). The environment also plays a relevant role moulding culture (D'Andrade 1995). Culture is largely the result of the recurring reactions of a group to face the problems posed by its environment, and it crystallizes elements of this experience. AT also considers these contexts. It focuses on the mutual dialectics between people and their physical and social environment: the environment shapes human actions and their execution, and is changed by these same actions. Hence, human acts cannot be analysed independently of their contexts, which become key in AT studies.

Finally, the study of culture cannot just consider isolated groups, as individuals are actually immersed in several interacting groups (Engeström 1987). This situation allows the exchange of artefacts and propagation of changes among linked cultures. AT considers that activities always exist interconnected in networks where different persons act and share some common background and artefacts. AT studies the basic principles governing these networks to know how changes spread across their levels and components. These contextualised networks are called here *social activities*.

A detailed comparison of AT and other paradigms to study culture is beyond the scope of this chapter. The interested reader can find more information in (Engeström 1987; Fuentes-Fernández et al. 2010; Leontiev 1978).

4.3 An Overview of Activity Theory

An *activity* (Leontiev 1978) is the minimal meaningful unit of analysis in AT. It is a transformation process driven by people's needs. An activity is potentially able to satisfy certain *objectives*, so people whose needs partially match with those *objectives* have in the activity a means to reach them. The satisfaction of the objectives is achieved when the activity produces its *outcomes* transforming *objects*. Any element used in this process is a *tool*. The active component that carries out the activity is the *subject*, who chooses among alternative *activities* following his/her own rationality. *Subjects* with a set of common social meanings and artefacts constitute a *community* (Engeström 1987), which represents the socio-historical context. Two bodies of social constructions mediate the relationships of communities in the activity: *rules* with the subject and the *division of labour* with the object. Both of them contain similar elements, such as knowledge, implicit assumptions or norms, but differ in their focus. The *division of labour* regards task specialisation in the community through aspects such as power relationships, goal decomposition or the assignment of responsibilities. On the contrary, *rules* are guides and constraints not targeted specifically to the activity but affecting it, such as group beliefs, country laws or scientific theories. These elements in the context of an activity constitute its *activity system*. All of them can be physical or mental, and AT considers both types with a unifying analysis. The traditional representation of an activity system (Engeström 1987) appears in Fig. 4.1.

An isolated activity is not enough to represent a complex social system. Activity systems always exist linked by shared elements. The execution of an activity produces outcomes that become the artefacts (e.g. subjects, tools or rules) needed

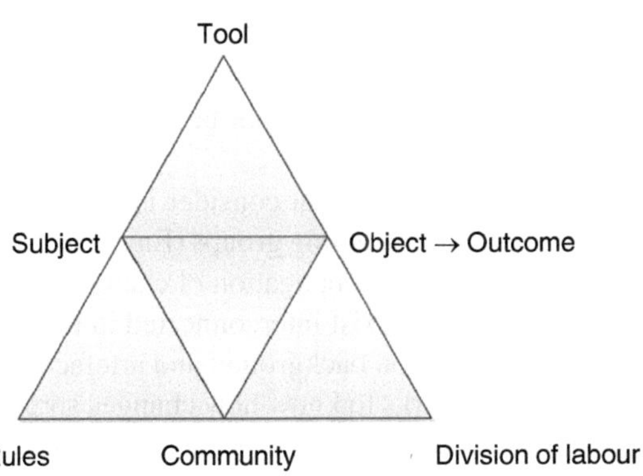

Fig. 4.1 AT depiction of an activity system

to execute other activities. When the activities in these networks share part of their social context (e.g. communities, rules or objectives), the networks constitute *social activities*.

AT also considers hierarchical decompositions. *Objectives* represent the key needs subjects consciously pursue; the intermediate *goals* identify alternatives or steps to achieve objectives; and the lower-level *states of the environment* describe specific contexts for task execution. Objectives, goals and states of the environment are achieved through *activities*, *actions* and *operations* respectively. In order to simplify the discussion, this presentation only considers *objectives* and *activities*.

Social activities evolve over time. Their elements modify their features (e.g. they grow up or suffer damages), and they appear and are destroyed. These changes produce conflicts intra or inter elements caused by competing goals or inadequacies between their features and purposes. AT names these conflicts *contradictions* (Engeström 1987). When contradictions appear, subjects face difficulties to satisfy their needs, as activities do not produce their outcomes with the required constraints. For this reason, subjects try to remove contradictions by changing the activity systems, for instance replacing some of their elements or creating new activities. These changes commonly generate new tensions that produce further evolution.

To illustrate these concepts, consider the example of a couple managing its marital problems (Swidler 2003). There is a high-level activity: *live together*. Its subjects are the two members of the couple, whose main objective is reaching *happiness* together. For this purpose, they work on their *lives* as the objects of the activity, trying to share thoughts, vital goals, leisure... They use different tools, such as their knowledge about relationships and the other, a restaurant for a romantic dinner, or making tasks at home. The related communities and their cultures frame the activity. This couple is American, suburbanite, middle-class, and in its mid-30s. Consequently, it belongs to several overlapped communities, such as Western people and mid-class couples. Each of these communities has some rules (or thoughts), like the popular wisdom that considers marital advisors (i.e. ministers and therapists) as experts in relationships, or choosing individual self-assertion when problems arise in couples. The couple undergoes a marital crisis: the current form of its activity *live together* no longer fully meets the objective *happiness*, producing a contradiction. This inadequacy takes the couple to look for new activities to fix it, like going to the advisor. This means a change in the network of activities related to their marriage. There is a new activity *visit the advisor* whose outcome *advice* is a tool to improve its activity *live together*. These activities are interconnected and working over the couple marriage, sometimes indirectly. They constitute the social activity of our case study.

4.4 A Framework for the Computational Analysis of Culture

AT was initially formulated for human interpretation and use, and not for computational analysis and simulation, so it needs to be formalised and adapted for this purpose. This implies the definition of computational models that satisfy

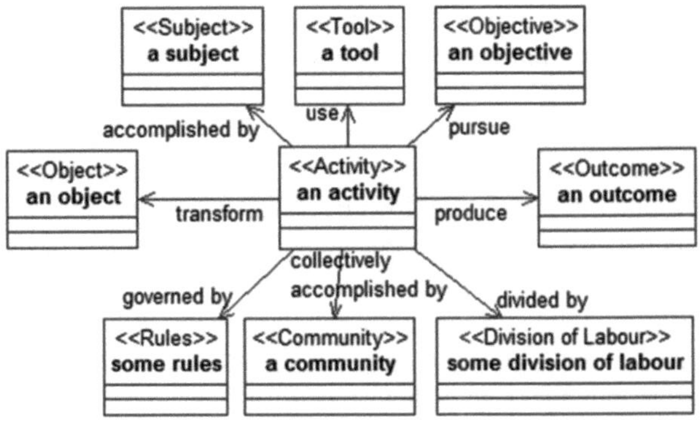

Fig. 4.2 UML-AT representation of an activity system

its principles. Our research has addressed these issues for Multi-Agent Systems (Fuentes-Fernández et al. 2009) and Requirements Engineering (Fuentes-Fernández et al. 2010). The work presented here discusses the application of these results for the analysis of culture. This application is based on a modelling framework whose main elements are: the modelling language used for societies and their cultures (see Sect. 4.4.1), its informal semantics based on the role of social activities in the application and construction of culture (see Sect. 4.4.2), and *social properties* to reuse fragments of specifications (see Sect. 4.4.3), being the representation of AT contradictions a key example of their use (see Sect. 4.4.4).

4.4.1 The UML-AT Modelling Language

Coming from Social Sciences, AT works describe their models, properties and results mainly using natural language complemented with arbitrary resources (e.g. pictures, video or recordings). These means are suitable when researchers and their audience share a background in Social Sciences: researchers can take advantage from all the expressive power of such variety of resources, while the readers' knowledge of the domain limits the potential problems of misunderstandings or ambiguity when using the information. However, this is not the case of computer experts collaborating in the development of simulations, so there is a high risk of misunderstandings (Drogoul et al. 2003). Moreover, these experts are concerned with the representation of information for automated processing, and thus used to modelling and programming languages.

Trying to fix some of the previous problems, we defined a modelling language that partly formalises the conceptual framework of AT. This language is called UML-AT (Fuentes-Fernández et al. 2007b) because it is a Unified Modelling Language (UML) profile (2010). It defines stereotypes for the main concepts and relationships of AT and their related constraints, such as the stereotypes that a relationship can link or its cardinality. Figure 4.2 shows the representation of the activity system in Fig. 4.1 using UML-AT.

UML-AT presents several differences with the traditional representations in AT. First, it makes explicit the relationships between concepts with primitives that effectively allow multiple cardinalities and networks of activities. AT highlights the relevance of the mediation relationships, but it demotes to text their discussion. The graphical representation of these relationships (see Fig. 4.1) relies on the positions of the linked concepts in the triangle. This representation is cumbersome when activity systems have multiple elements of each type or when representing elements shared among several activity systems. Second, UML-AT incorporates additional concepts that facilitate the specification of social systems in contexts related to computational systems. Some of these additions are the concept of artefact and the relationships of inheritance, contribution, decomposition and change of role.

The concept of *artefact* represents a generic element that participates in an activity system and that can be refined to any of the other roles. This allows specifying requirements for the execution of an activity when it is not relevant, or it is still unknown, the specific role the concept plays in the system.

In order to facilitate the incremental specification of systems, UML-AT includes inheritance relationships between artefacts. Inheritance permits to define abstraction layers, starting with generic concepts and successively refining them to add specific properties. A sub-concept inherits all the properties and participates in all the relationships of its super-concepts, and it can also constrain or extend them.

Contribution relationships describe how artefacts influence the options of others to achieve their purpose. Examples of them are conflicts between objectives, the use of a tool damaging another artefact, or knowledge that must be available to interpret properly certain procedures. These relationships extend those of the language i* (Yu et al. 1997) and include, for instance, *guarantee*, *essential*, *impede* or *contribute negatively*.

Decomposition relationships are concise means to indicate relationships whole-part without introducing activities with no real meaning. For instance, they are used to describe the refinement of objectives and activities.

Finally, a *change of role* relationship accounts for the different roles that the same artefact can play in different activity systems. Social systems are networks of activity systems linked by shared artefacts, and these artefacts do not necessarily have the same purpose in all the systems. A typical example is that the object or subject of an activity system is the result (i.e. outcome) of other systems.

The construction of models with UML-AT follows the activities depicted in Fig. 4.3:

1. *Is the model complete enough?* Modellers create their specifications performing cycles where they add or refine parts of models. When a cycle has finished, they consider whether their specifications reflect accurately and with enough detail the social system under study for the problem at hand. If the answer is affirmative, the process finishes, but if not, they need to perform new cycles.
2. *Identify activities.* The basis of the AT analysis is the acts people carry out. Initially, modellers should discover those activities that constitute identifiable units of action with a beginning and ending.

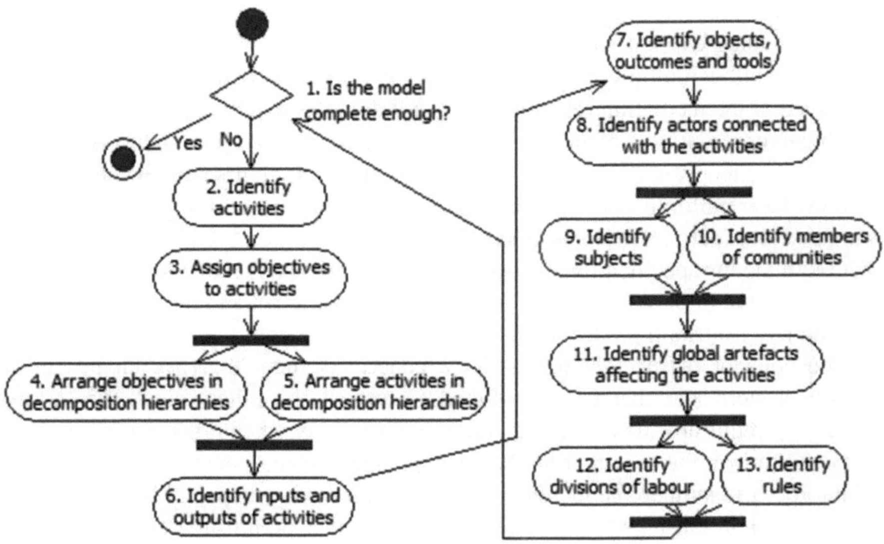

Fig. 4.3 Modelling process: activities and partial ordering

3. *Assign objectives to activities*. Subjects execute activities for a given purpose. An activity can have many motives, that can be pursued under different constraints.
4. *Arrange objectives in decomposition hierarchies*. The objectives identified in certain environment are usually related through contribution relationships. These relationships can be difficult to grasp, so the study in activity 5 of their related activities can help to clarify these hierarchies.
5. *Arrange activities in decomposition hierarchies*. Activities are related through decomposition and producer-consumer relationships. The first type of relationship focuses this step. These decomposition hierarchies are not always evident. The analysis of the objectives related with the activities can help here.
6. *Identify inputs and outputs of activities*. Activities are transformation processes that require certain artefacts as input and produce or affect certain artefacts.
7. *Identify objects, outcomes and tools*. For an *activity*, the results of its execution are its *outcomes*. These include artefacts generated, modified or destroyed by the *activity*. Regarding the inputs of the *activity*, there is a neat distinction between *objects* and *tools*. *Objects* are transformed to constitute the *outcomes*, for instance because they are their original parts. *Tools* are used to perform this transformation. Modellers must consider that from the AT perspective there is no distinction at this level between physical and mental artefacts. For instance, a tool can be a notepad or a phone, but also a plan or knowledge about languages.
8. *Identify actors connected with the activities*. This activity considers those actors, i.e. active elements, who contribute to the execution of an *activity*. This

participation is in a broad sense, covering from the direct execution of the activity to people providing required elements or influencing the execution.

9. *Identify subjects*. *Subjects* are the actors that actually perform the *activities*.
10. *Identify members of communities*. *Communities* are groups of people affecting the activity. This influence comes in different ways, such as shared development of artefacts, participation in the historical development of the activity, generation of the knowledge required to carry it out, or education of its subjects.
11. *Identify global artefacts affecting the activities*. The execution of activities largely depends on artefacts that do not belong to any particular subject but to communities. This is the case of the norms and knowledge of a civilization. These artefacts crystallise the experience of a group performing certain *activities*.
12. *Identify divisions of labour*. The global artefacts that are targeted to a specific activity and its related domain constitute its *division of labour*. A typical example of it is the organization of a group of people to perform some shared work, including the roles they play in the group and the responsibilities of each of them, or the planned workflows.
13. *Identify rules*. Rules are global artefacts that groups develop for a broader domain than a set of related activities. They are intended to affect several aspects of the group activities. For instance, laws, religious and political points of view, scientific theories, or shared interests belong to this group.

Some remarks must be done regarding the previous modelling process. First, the social activities used to model culture emerge in the process through the artefacts shared among the different activity systems. Following AT principles, the basic analysis unit is the individual *activity*. Second, the process has been represented as mainly sequential for simplicity reasons. However, its practical application leads to a more interleaved application of the different activities. For instance, modellers can start by identifying the subjects in activity 9, then their objectives in activity 4, and later use these to discover the related activities in step 3. Third, the designation of roles for the different artefacts in activity systems depends largely on the focus in the analysis. For instance, when considering social trends, e.g. dress codes or meeting protocols, it is usually enough modelling them as rules or division of labour with an indirect influence in the subjects' behaviour; however, they can become objects or tools in formal contexts, for trend aware people, or when studying them in academic works.

Our use of UML-AT to study culture through simulation achieves two main goals. First, it serves to produce an intermediate language that could be useful for both social researchers and engineers (Fuentes-Fernández et al. 2010). The first group interprets UML-AT specifications based on its knowledge about AT, and the later uses its expertise in UML to grasp part of the intuitive meaning of AT concepts in UML-AT. There is still a gap in the details of the specifications: social researchers use mainly natural language for them, while engineers choose formal or semi-formal languages. Nonetheless, the use of a common abstract specification language, such as UML-AT, largely reduces this gap when compared with traditional approaches (Drogoul et al. 2003). Second, a modelling language with a formal definition is a suitable basis for a Model-Driven Development (MDD) (France et al. 2007) of simulations. MDD is based on the specification of software systems with models conforming to modelling languages, and the use of semi-automated transformations for a gradual refinement of those models from requirements to code. Thus, MDD supports the kind of incremental and documented development of simulations from the abstract social models advocated in literature (Edmonds 2003).

4.4.2 Discussion on Semantics

The use of a modelling language such as UML-AT to specify social systems does not guarantee that modellers make a correct use of it. Its formal definition describes its syntax, but not its intended semantics. This situation is quite common in Computer Science, see for instance the examples of UML (2010) or languages in MDD (France et al. 2007). Following the activities presented in the previous section reduces the chances of building incorrect models, but still, an introduction to concrete semantics is needed.

The informal semantics presented in this section for UML-AT can be understood as the interpretation of the behaviour over time of activity systems. Our work adopts a simple producer-consumer model. Given some specifications and an execution of them, their artefacts can be available or not. An artefact is available if it was an initial element in the system or it was produced by an activity as an outcome, and it was not destroyed later by another activity. When all the artefacts of an activity system but the outcomes are available, its subject can decide to trigger the execution of the activity. The model does not prescribe any specific rationality for subjects when choosing among several potential activities to fire. Multiple subjects can choose to execute activities at the same time. After the execution of an activity, its outcomes are changed as the specifications describe.

The result of this execution is a tree of states, similar to those in branching time temporal logics (Clarke and Grumberg 2000). A state describes what artefacts of the specifications exist at a given moment and the assignment of values to their attributes. Successive states in a branch of the tree represent successive moments, and alternative branches different choices about the execution of activities.

UML-AT and its semantics are the basis for the verification of properties over specifications. UML-AT diagrams describe both specifications and properties. The attributes of their components (e.g. name or type) can be either variables or constant values. The basic verification process uses pattern matching to check static aspects (as those seen in Sect. 4.5). This process looks in the specification of the system for sets of artefacts that match the specification of a property. The match means that a set of artefacts has the same entities, relationships and attributes as the property, where constant values only match with themselves and variables match with any value. The verification of properties about the dynamic behaviour of systems requires additional techniques, such as simulation and model checking (Clarke and Grumberg 2000). In both cases, the run of the simulation is seen as a tree of states. Model checking is only suitable when the state space is limited and it can be exhaustively explored. Given the usual complexity of social systems, a user-guided interactive exploration of the state space through simulation is more common. Pattern matching of properties over states can be applied in both cases, but the expression of complex temporal properties usually requires the use of some kind of temporal logics too, and it is more common with model checking.

This verification approach is still limited given the expressive power of UML-AT. It only distinguishes the key roles and relationships of AT. For instance, deciding the actual meaning of a rule and its effect in the execution of an activity is largely a task for experts. The models just allow an abstract modelling and organization of the information, and the verification process supports looking for properties based on structural relationships of specifications. However, these benefits already bring an important effort saving for experts.

4.4.3 Social Properties

The specification of social systems to analyse their culture is a demanding task. In order to reduce this effort, our work tries to take advantage of the similarities between different scenarios. Many of the aspects that describe a culture are relevant in different settings. For instance, the prevalent perspective in a country about relationships permeates its laws, couple interactions, and the activities of people getting to know others. Some cultural aspects are also similar in different groups. For instance, Western societies put on high relevance the individual, and this influences the rights they assured to persons and the political organization, though there are differences in their actual implementation. The concept of *social property* aims at describing these reusable aspects of specifications.

A *social property* is a specification of an aspect of a social system that is recurrent in different scenarios. The use of these properties intends to satisfy several purposes: to document a social aspect; to act as a communication device between social experts and engineers; and to support the development of systems, e.g. simulations, workflow management systems, or social network applications. To meet all these objectives, social properties are described with the structure represented in Fig. 4.4.

A *social property* includes a unique *identifier* and a *description*. The description explains its general meaning, the type of context where it can be useful and how it can affect the specification of other parts of the social system.

The detailed specification of the property is organized around *settings*. A *setting* describes a property in a specific context of application, i.e. a kind of social group and environment. For instance, a social property can describe a structure of collaborative peers, and its settings account for the differences between organizations with peers arranged in the same place or distributed and using communication devices.

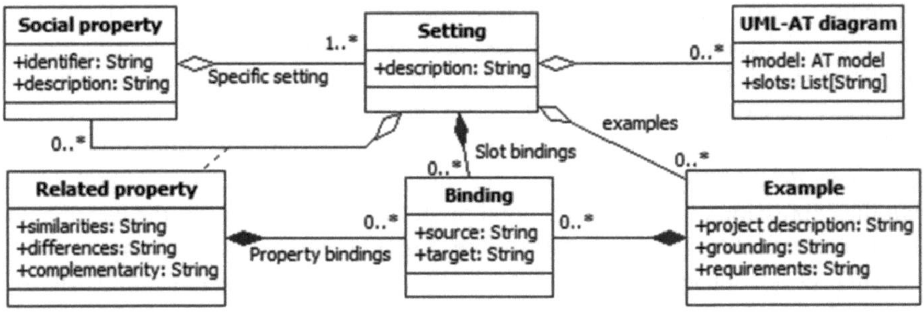

Fig. 4.4 Structure to describe *social properties*

The description of a *setting* includes a textual *description* and several *UML-AT diagrams* with their model and list of slots. The list of *slots* indicates which elements of the diagram are constants (only depend on the setting) and which variables (depending on the system under study). In order to facilitate the understanding of settings, they can have *examples* of use. These examples show how a setting appears in a study and the meaning of its different elements in that context.

The *related properties* link settings and properties whose contexts or artefacts are related. For instance, a setting of the peer organization can be related to other settings that describe normative frameworks for these organizations, or to other alternative social structures. The information included in the *related property* discusses their similarities and differences.

The last element of the description is the *bindings*. These are pairs of names that link a setting and its examples or a setting and its related properties. In the first case, they indicate the specific values that the variables of the setting take when grounded in the example. In the second case, they indicate potential relationships between concepts. For instance, the outcome of an activity is the tool in a related property.

An important remark about social properties is that they facilitate the kind of MDD approach (France et al. 2007) that was mentioned in Sect. 4.4.1. Being reusable modelling components, they are refined in several projects with different requirements. The result is a set of properties with multiple settings at different levels of detail (e.g. abstract requirements or platform-oriented), with different contexts (e.g. cultures or countries), and targeted to different transformations (e.g. simulation platforms or documentation). Such catalogue of ready-to-use properties for projects assists in the specification with predefined parts of models and suggests potential alternatives for model refinement.

4.4.4 Contradictions

As introduced in Sect. 4.3, contradictions are situations that drive subjects to evolve activity systems over time. Hence, they are also needed to study the change of

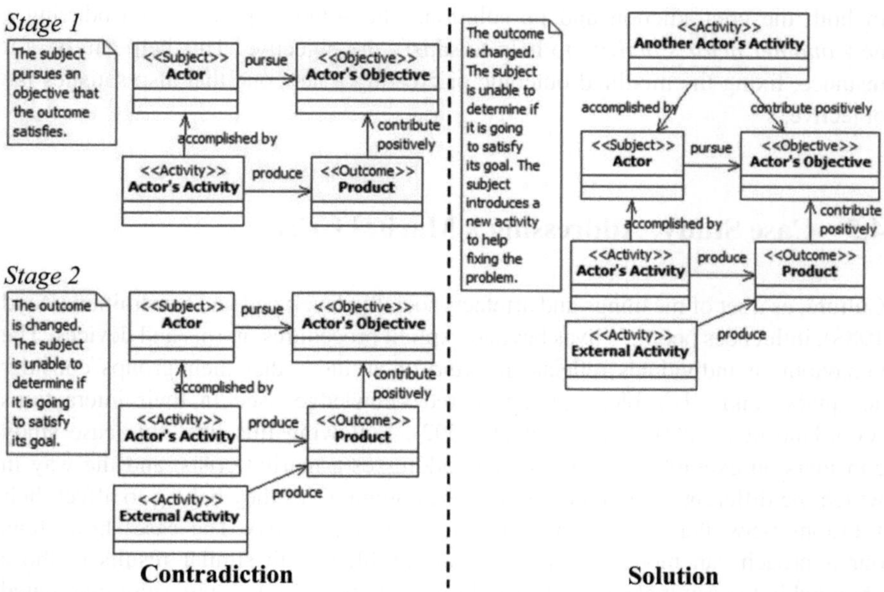

Fig. 4.5 Social property of the *Need State* contradiction and a related solution

cultural aspects in a society. AT literature identifies several standard contradictions. Some of them have already been adapted to UML-AT (Fuentes-Fernández et al. 2007a), e.g. the *Conflict Producer-User*, the *Double Bind*, or the *Need State*, which will be used later on in the case study.

The *Need State* contradiction (Bratus et al. 1983) accounts for states where a subject relies on the product of a given activity to satisfy certain needs. Changes in the environment can affect that product and render it unsuitable for that purpose. For instance, the couple talks to keep its confidence, but when some close friends point out that this talk is being meaningless, it loses its power to provide that benefit. This lack of alignment between the objective, the activity and the modified outcome causes the contradiction.

Figure 4.5 shows on the left part of the specification that contradiction. In relation to Fig. 4.4, it shows the *UML-AT diagram* of a *setting* of this *social property*. Stage 1 corresponds to the original situation without conflicts where the *actor* produces the *product* that satisfies the objective. Stage 2 describes the situation where the potential contradiction could appear. The change in the environment represented by the *external activity* affects the *product*, so it can become unsuitable for its purpose. Some *examples* of this property could be extracted from the case study in Sect. 4.5.

The representation of social properties also considers their *related properties*. For a contradiction, these can be potential solutions for it. *Related properties* usually just point out to their corresponding social properties. In this case, Fig. 4.5 draws together the contradiction on the left and a solution on the right to facilitate the presentation. *Bindings* are represented assigning the same name to linked concepts in both the contradiction and its solution. The solution proposes introducing a new *another actor's activity* to help to satisfy the objective. This help can be, for instance, fixing the modified outcome or creating a new one that also satisfies the objective.

4.5 Case Study: Addressing a Marital Crisis

Culture, as a set of meanings and artefacts socially developed and transmitted (Boyd 1988), influences people's lives beyond explicit procedures, norms and devices. The behaviour of individuals reflects the kind of attitudes that their groups consider acceptable and advisable, and the shared knowledge used in their interactions (Giordano et al. 2006; Simon et al. 1992). Following this line, this case study considers an example of how a couple addresses a marital crisis, and the way in which the different cultural contexts of the communities they belong to affect their solutions ((Swidler 2003) in part I: *Culture's confusions*). The case shows how our approach can model and analyse this problem with similar results to those obtained in the original study with traditional approaches, but with semi-automated techniques that reduce the researcher's effort. The networks of activities at a given moment constitute the social activities used in this study.

The case study presents Emily and Frank. They were a typical North-American middle-class couple in their 30s at the end of the 1990s. After several years of happy marriage, they underwent a serious marital crisis. They looked for advice to fix their problems in therapists and ministers. The prevailing culture of their society took these advisors to recommend them doing what was better for them as individuals, looking for their self-development and trying to express themselves. However, Emily and Frank were not completely happy with this solution, as they really wanted to stay together. They had also participated in the "Marriage Encounter" movement, where the focus was on commitment to one's marriage and shared feelings with the partner. Nevertheless, the plain commitment was not enough to fix their problems. At the end, they built a self-development attitude but inside their couple, saving the relationship.

The analysis of this case with our approach includes one view for the basic social contexts and four more that reflect different stages in the couple, all of them expressed with UML-AT. These stages are: the pre-conflict stage, previous to the crisis; the conflict stage when the couple felt no longer happy together; the failing solution stage when they talked with the advisors; and the reworked solution stage when they reoriented counsels and attitudes.

First, the analysis requires considering the information about the social contexts of the problem. Figure 4.6 shows these. There were two communities involved in the problem, the *modern Western society* and the *Marriage Encounter movement*. The first one gathers the results of the historical development of Western societies, where individuals are at the core of thought instead of the group. Moreover, for the last decades, there is a prevailing attitude to consider personal relationships

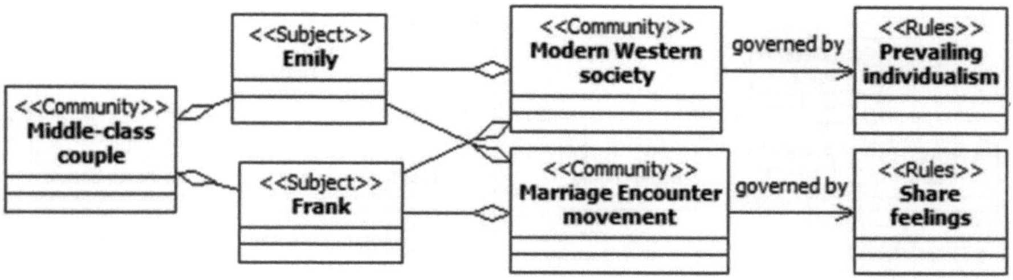

Fig. 4.6 Cultural contexts of the couple

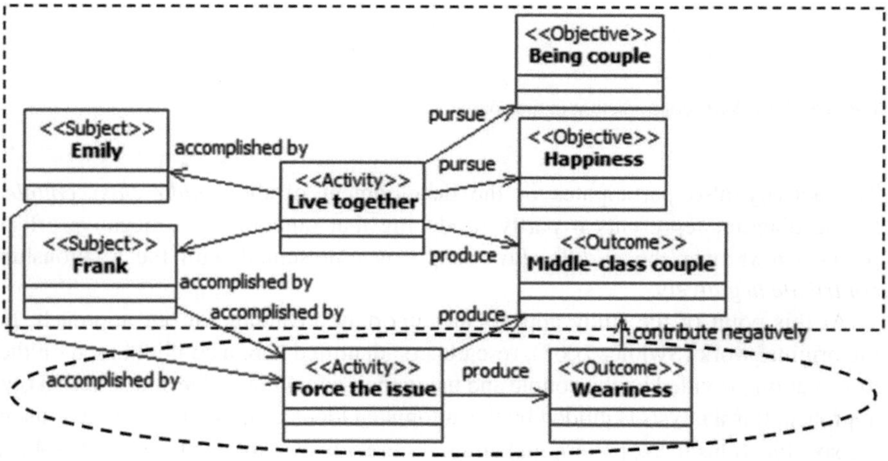

Fig. 4.7 Pre-conflict and conflict stages. Pre-conflict elements appear inside the dotted rectangle. New elements in the conflict stage appear inside the dotted ellipse

through individual perspectives, with concepts such as autonomy, self-development and self-assurance. This framework is considered with the rules of *prevailing individualism*. At the same time, this couple participated in a movement that promoted the relevance of marriage and made *share feelings* a key norm to improve relationships.

In the beginning of the case study, Emily and Frank constituted the happy couple shown in Fig. 4.7. They were the subjects of a shared activity *live together* whose product was their *middle-class couple*. This result satisfied for them two objectives: reaching *happiness* and *being couple*.

The problem arose when they felt that the couple was not working as seamlessly as before. They were forcing themselves to behave as if nothing had changed, but they did not feel as that. Figure 4.7 summarizes this situation adding some new elements for the second stage. The activity *force the issue* represents the new effortful tasks. It has the specific outcome *weariness*, which compiles the negative feelings that the continuous effort and dissatisfaction produced in the couple.

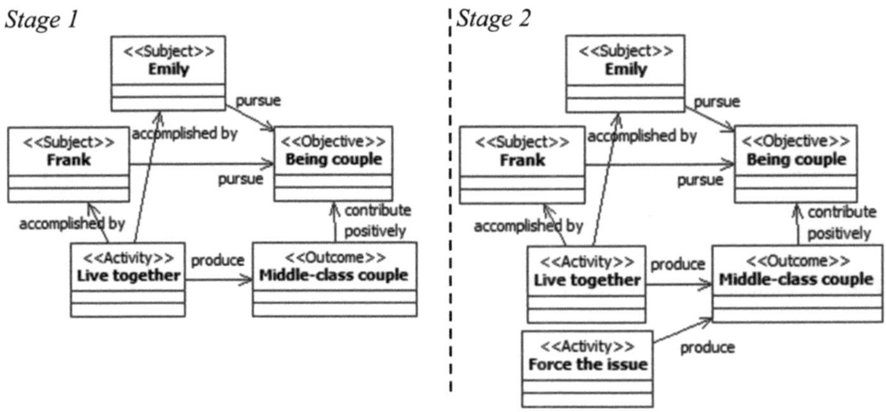

Fig. 4.8 *Need State* contradiction in the couple

The activity also participates in the development of the *middle-class couple*, so the diagram represents it partly producing that outcome. The negative effect of *weariness* over the *middle-class couple* is represented with the relationship *contribute negatively*.

At this point of the study appears an important advantage of our approach. In the original work (Swidler 2003), researchers identified here a conflict based on the information provided by the couple and using their own knowledge and skills. In our approach, this analysis is guided by the automated identification of contradictions in the specifications using pattern matching. As previously explained (see Sect. 4.4.2), this process looks in the specifications for subgraphs with the same structure as the property. For instance, Figs. 4.6 and 4.7 have a correspondence with the *Need State* contradiction in Fig. 4.5. Figure 4.8 shows this mapping by replacing the abstract elements in Fig. 4.5 with the actual ones from the case study. It indicates that the situation of the couple could contain such contradiction and provides an interpretation for it: the outcome *middle-class couple* is no longer only the product of the activity *live together*, as a new activity *force the issue* also affects it. These additional modifications may render the product unsuitable to satisfy the objective *being couple*. Note that the matching does not need to use the information about the negative contribution of *weariness* to the *middle-class couple*. An expert later evaluates this identification to check whether it really makes sense in the current context.

The study continues showing in Fig. 4.9 the results the couple got visiting the marital advisors. A third subject, the *advisor*, also participated in this activity. The prevailing social context, i.e. *modern Western society*, moulded this activity using its rules of *prevailing individualism*. Consequently, the outcome of these sessions, represented by *self-development attitudes*, focused on people being almost uniquely concerned and responsible for their own life. From the perspective of the couple,

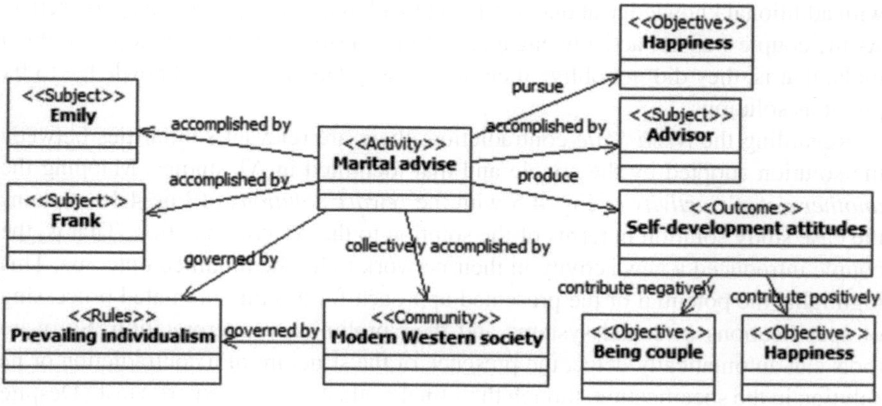

Fig. 4.9 The failing solution stage

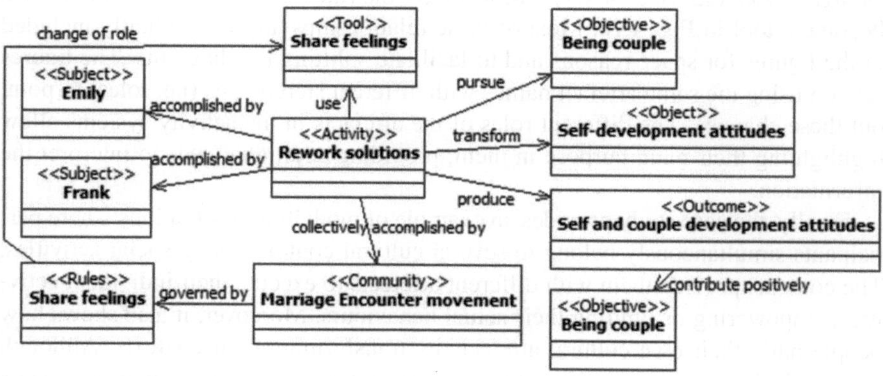

Fig. 4.10 The reworked solution stage

this kind of advice could be positive for their individual *happiness*, but not for their common objective of *being couple*. These impressions are represented with contribution relationships.

Finally, Emily and Frank reworked the advisor's counsels to produce their own solution. The couple felt they effectively needed more individual self-assertion, not only as a way to evolve as individuals, but to foster the elements they could share and develop together as a couple. For this process, they used as a tool the knowledge about *share feelings* obtained in their experience with the *Marriage Encounter movement*. Figure 4.10 summarizes these results. The *self-development attitudes* from Fig. 4.9 are here the object of a new activity *rework solutions*. Its outcome is the adapted *self and couple development attitudes* that Emily and Frank found useful to achieve their objective of *being couple* and happy together. In this context, the *Marriage Encounter movement* was a community that fed the couple with additional knowledge about how to deal with relationships from its perspective. As the couple was not actively engaged with the movement, its rules were for them tools, that is, they did not oblige them in any way but constituted knowledge to fix previous solutions.

Regarding the *Need State* contradiction, there are relevant similarities between the solution adopted by the couple and that identified in AT studies. Mapping the *another actor's activity* of Fig. 4.5 with the *rework solutions* of Fig. 4.10 explains the case study solution in terms of the solution to the AT contradiction. That is, the couple introduced a new activity in their network to fix the modified outcome. This highlights the potential of the presented approach for a semi-automated processing of specifications of social systems and their preliminary interpretation. Software tools can automatically detect the presence of the structure of a contradiction or its solution in the specification, though their final evaluation is an expert's task. Despite of this limitation, this approach brings an important effort saving for experts.

A remark about the different diagrams is the pervasive use of the relationship *change of role* (see Sect. 4.4.1). For instance, the rules *share feelings* from Fig. 4.6 become a tool in Fig. 4.10. Most of these relationships are not explicitly included in the figures for space reasons and to facilitate splitting the diagrams. The figures rely on using the same artefact names with different stereotypes (i.e. roles) to point out these changes. The different roles of the artefacts in the activity systems allow highlighting their main purpose in them, providing additional cues to interpret the information.

Finally, the case study provides an example of modelling in situations where participants simultaneously belong to several cultural contexts using social activities. The contexts provide them with different artefacts to execute their individual activities, empowering or limiting their actual behaviours. Moreover, it also shows how people build their own cultural artefacts by transforming their contexts. Although the couple does not constitute a group with its own culture, their adaptation process in the social activity shows the emergence of new artefacts that when transmitted to a wider group could constitute the seed of new elements of culture.

4.6 Conclusions

This chapter has shown the use of a modelling framework based on AT to analyse the influence of culture through social activities. It conceives culture as an organised set of socially built and transmitted artefacts, which can be either physical or mental. Following AT, its study focuses on activities as the analysis units where those artefacts are developed and used. The social dimension of culture is represented with social activities, i.e. networks of interconnected individual activities whose contexts share the elements that constitute the culture of the involved communities.

The framework has an infrastructure built around the modelling language UML-AT, which formalises the AT conceptual framework. This language is used to specify social systems and their properties. The effort invested in modelling and interpreting information is reused with social properties. These social properties are fragments of models with complete descriptions of their meaning, and connected to other properties through relationships that point out semantic links.

Our work includes a guideline that indicates how to identify, in a given domain, the concepts relevant for the specification with AT. There is also a simple dynamic model of behaviour that specifies the evolution of activity systems over time.

The previous elements enable a semi-automated analysis of the properties of specifications based on pattern matching. Software tools identify in the specifications subgraphs with the same structure of the considered properties. Though experts must later analyse its results, this provides a first filter of the information. When dealing with social properties, the matching also provides the basis for a preliminary interpretation of the information according to the definition of those properties.

The features of the framework have been illustrated with a case study about couple relationships. It models with social activities several aspects of the original problem. The use of social properties automates the preliminary interpretation of the specifications. It shows how the couple solved its contradictions through the generation of new shared artefacts from its cultural contexts. These artefacts become potential culture components when transmitted to the couple's communities.

The framework is still ongoing work. There are several open issues to address. First, the framework should provide more detailed guidelines and metrics to facilitate the modelling process and identify possible variation points. In the case of social activities, it is particularly difficult finding the elements of the shared contexts, as they are frequently implicit knowledge and norms. Our current work also tries to measure the relationships between certain metrics (e.g. size and connectivity of social activities, and appearance of some social properties) and the quality of models perceived by experts. Second, UML-AT needs to be extended with additional primitives, mainly for dynamic and normative aspects. Now, modellers only indicate the temporal sequence of activities. Extra primitives could indicate duration, concurrency and feasibility. In the same way, rules and divisions of labour only specify simple conditions about the presence or absence of artefacts. The integration of features of modal logics in UML-AT is here the main line of work. Finally, we think that a rich catalogue of social properties is a key element to carry out an effective use of automated analysis for social systems. Our research is working on available literature to extend this catalogue.

Acknowledgements This work has been done in the context of the project "Social Ambient Assisting Living – Methods (SociAAL)", supported by Spanish Council for Economy and Competitiveness, with grant TIN2011-28335-C02-01. Also, we acknowledge support from the Programa de Creación y Consolidación de Grupos de Investigación GR35/10-A.

References

Boyd, R., Richerson, P. J.: Culture and the Evolutionary Process. University of Chicago Press (1988).

Bratus, B. S., Lishin, O. V.: Laws of the development of activity and problems in the psychological and pedagogical shaping of the personality. Soviet Psychology XXI, 38–50 (1983).

Clarke, E. M., Grumberg, O., Peled, D. A.: Model Checking. The MIT Press (2000).

D'Andrade, R. G.: The Development of Cognitive Anthropology. Cambridge University Press (1995).

Drogoul, A., Vanbergue, D., Meurisse, T.: Multi-Agent Based Simulation: Where Are the Agents? In: Simão Sichman, J., Bousquet, F., Davidsson, P. (eds.) Multi-Agent-Based Simulation II. LNAI, vol. 2581, pp. 1–15. Springer, Heidelberg (2003).

Edmonds, B.: Towards an Ideal Social Simulation Language. In: Simão Sichman, J., Bousquet, F., Davidsson, P. (eds.) Multi-Agent-Based Simulation II. LNAI, vol. 2581, pp. 105–124. Springer, Heidelberg (2003).

Engeström, Y.: Learning by expanding: An activity-theoretical approach to developmental research. Orienta-Konsultit (1987).

France, R., Rumpe, B.: Model-driven Development of Complex Software: A Research Roadmap. In: 2007 Future of Software Engineering (FOSE 2007), pp. 37–54. IEEE Computer Society (2007).

Fuentes-Fernández, R., Gómez-Sanz, J. J., Pavón, J.: Managing Contradictions in Multi-Agent Systems. IEICE Transactions on Information and Systems E90-D(8), 1243–1250 (2007).

Fuentes-Fernández, R., Gómez-Sanz, J. J., Pavón, J.: Model integration in agent-oriented development. International Journal of Agent-Oriented Software Engineering 1(1), 2–17 (2007).

Fuentes-Fernández, R., Gómez-Sanz, J. J., Pavón, J.: Requirements Elicitation and Analysis of Multiagent Systems Using Activity Theory. IEEE Transactions on Systems, Man and Cybernetics, Part A: Systems and Humans 39(2), 282–298 (2009).

Fuentes-Fernández, R., Gómez-Sanz, J. J., Pavón, J.: Understanding the human context in requirements elicitation. Requirements Engineering 15(3), 267–283 (2010).

Giordano, P. C., Longmore, M. A., Manning, W. D.: Gender and the Meanings of Adolescent Romantic Relationships: A Focus on Boys. American Sociological Review 71(2), 260–287 (2006).

Leontiev, A. N.: Activity, Consciousness, and Personality. Prentice-Hall (1978).

OMG: OMG Unified Modeling Language (OMG UML), Superstructure, Version 2.3. May 2010, http://www.omg.org (2010).

Railsback, S. F., Lytinen, S. L., Jackson, S. K.: Agent-based Simulation Platforms – Review and Development Recommendations. Simulation 82(9), 609–623 (2006).

Shore, B.: Culture in Mind: Cognition, Culture, and the Problem of Meaning. Oxford University Press (1996).

Simon, R. W., Eder, D., Evans, C.: The Development of Feeling Norms Underlying Romantic Love Among Adolescent Females. Social Psychology Quarterly 55(1), 29–46 (1992).

Swidler, A.: Talk of Love: How Culture Matters. University of Chicago Press (2003).

Vygotsky, L. S.: Mind and Society. Harvard University Press (1978).

Yu, E.: Towards Modelling and Reasoning Support for Early-Phase Requirements Engineering. In: 3rd IEEE International Symposium on Requirements Engineering (RE'97), pp. 226–235. IEEE Press (1997).

Part II
Group and Organisation Culture

Chapter 5
Cultural Integration and Differentiation in Groups and Organizations

Michael Mäs, Andreas Flache, and James A. Kitts

5.1 Introduction

Extensive research has documented the importance of organizational culture, but we are only beginning to understand the processes by which organizational cultures emerge, persist, and sometimes change or split into subcultures. Organizational cultures often prove to be remarkably stable, despite membership turnover, change of leaders, shifting social networks, or disruptive external forces. Enriching our understanding of the basic dynamics of organizational culture will foster theoretical advances with important practical implications, especially in preparing for challenges such as organizational change, growth, or merger.

To provide a rigorous microfoundation, we focus here on the dynamics of cultural influence in a simple, stylized model that allows us to generate testable predictions about the conditions of cultural consensus, cultural diversity, and polarization of cultures in organizations. Our analyses focus on the effects of "social differentiation", the tendency of individuals to adjust their opinions and values in order to increase differences to others. Social differentiation appears to be a critical assumption in models that seek to explain cultural diversity and has been supported by empirical research. However, we show that existing models are based on two different conceptualizations of social differentiation. Using computational

M. Mäs (✉)
Chair of Sociology, in Particular of Modeling and Simulation, ETH Zürich, Clausiussstrasse 50, 8092 Zürich, Switzerland
e-mail: mmaes@ethz.ch

A. Flache
Department of Sociology/ICS, University of Groningen, Grote Rozenstraat 31, 9712 TG Groningen, The Netherlands

J.A. Kitts
Computational Social Science Institute, Department of Sociology, University of Massachusetts, 200 Hicks Way, Amherst, MA 01003-9277, USA

V. Dignum and F. Dignum (eds.), *Perspectives on Culture and Agent-based Simulations*, Studies in the Philosophy of Sociality 3, DOI 10.1007/978-3-319-01952-9_5,

experiments, we demonstrate that these conceptualizations imply critically different patterns of polarization, radicalization, and factionalism. In addition, they generate cultural diversity under different initial conditions.

Most relevant research has employed formal theory to account for the emergence and persistence of cultural groups, showing how a population of agents with arbitrary opinions and social relations may over time develop a coherent collective culture. This work has overwhelmingly built on one of the starkest regularities in the social world: the tendency of social ties to connect individuals who are similar in attributes, attitudes, or behaviors. This observed lawlike regularity of differential attraction or homophily has inspired prominent "first principles" for models of local cultural emergence. First is the tendency for actors to build positive ties to interaction partners who are similar to themselves (Homans 1950). Second is social influence, the tendency for common attitudes or behaviors to diffuse among friends and other close relations (Festinger et al. 1950). This combination of differential attraction and influence creates a self-reinforcing dynamic in which similarity increases conformity between interaction partners and conformity increases similarity of interaction partners. Such positive feedback leads to a local homogenization that some have presented as an explanation for the emergence of "cultural norms" (Latane 2000) in social networks. Furthermore, such models have been used to understand the maintenance and stability of culture in organizations (Carley 1991; Harrison and Carroll 2002; Kitts and Trowbridge 2007) as well as the integration of multiple cultures, such as following a merger of two organizations (Carroll and Harrison 2002; Harrison and Carroll 2006).

Although the core dynamics of homophilous choice and conformity have received much empirical support, and they provide a convincing account for cases of cultural integration and homogeneity, they leave us instead with the opposite puzzle of explaining cultural diversity in densely connected groups. If homophilous attraction and conformity are such general forces, how may we ever explain the maintenance of distinct cultural subgroups (Bednar et al. 2010; Centola et al. 2007; Macy et al. 2003; Mark 1998) in contact with one another? In fact, it has been proven (Abelson 1964; Harary 1959) that positive influence operating on a fully connected graph (where each actor is connected to each other by at least one influence path) will under a broad range of conditions eventually result in a 'monoculture' where all individuals have the same opinions or attitudes. These models fail to explain why social groups and organizations often harbor a diversity of views, given that formal and informal networks are almost guaranteed to be connected and are often dense.

The most intuitive explanations for diversity posit exogenous factors that hamper cultural convergence or even create diversity. These "top down" accounts assume, for instance, that physical barriers, social and political cleavages, or boundaries between divisions of an organization somehow prevent social influence from flowing freely throughout the population (Parisi and Cecconi 2003). It has been similarly shown that conflicting political parties or media may exert influence on individuals' cultural attributes and interfere with cultural convergence (Shibanai et al. 2001).

In contrast to approaches that rely on exogenous barriers or influences, research has also shown that cultural diversity can result from "bottom up" self-organization within a population of agents. Applying the principle of homophily to an extreme case, some scholars (Axelrod 1997; Carley 1991; Mark 1998) assume that if two actors have disjoint cultures (share nothing in common), they then have zero propensity to interact with one another, creating a cultural boundary that operates like a geographic boundary. These models are then able to generate persistent diversity. In this case, the same local convergence that would lead to homogenization on a connected influence network can actually lead a network to disintegrate into disconnected components, where local influence paradoxically maintains cultural differences rather than erasing them. Once the members of two cultural subgroups have become too dissimilar to influence one another, their cultures evolve along divergent paths. This type of model thus incorporates both tendencies that are evident in cultural dynamics – on the one hand, the drive toward uniformity within local relations, and on the other, the persistence of diversity in the greater population. While much of this work has modeled opinion scales as discrete, other studies combined homophily with continuous opinion scales (Harrison and Carroll 2002; Deffuant et al. 2000; Hegselmann and Krause 2002). These so called "bounded confidence models" showed that global diversity does not depend on the assumption that opinions are discrete, so long as influence can only occur between individuals who are sufficiently similar.

Further research has shown the bottom-up theories of cultural diversity to be extremely fragile. Recent work (Klemm et al. 2003; De Sanctis and Galla 2009) relaxed the assumption that cultural traits are entirely determined by influence from neighbors and allowed a small probability of random perturbation of cultural traits. If this noise is sufficiently low, occasional overlap between distinct cultures due to random distortions leads to the eventual collapse of cultural diversity. But if noise is sufficiently high, mutation is introduced faster than conformity can reduce it, leading to cultural turbulence that precludes the formation of stable subcultures. The window of conditions that allows cultural diversity in between these two regimes is exceedingly small and all but vanishes in larger populations. A second problem with these explanations of self-organized cultural diversity is that they rely on the assumption that cultural influence is entirely precluded when interacting agents are too dissimilar. Even slight influence between agents who are highly dissimilar is sufficient to eliminate cultural diversity based on homophily and conformity alone, a result that has been obtained for models with discrete as well as with continuous opinion spaces (Flache and Macy 2011; Mäs et al. 2010).

We focus on two solutions to the problem of self-organized cultural diversity, which were inspired by theories of social differentiation in classical sociology (Bourdieu 1984; Durkheim 1997; Elias 1969) and social psychology (Brewer 1991; Snyder and Fromkin 1980; Tajfel and Turner 1986).

The first approach invokes "distancing" as a key driving force of social differentiation, drawing on balance theory (Heider 1967) and cognitive consistency theories (Festinger 1957) from social psychology. Just as homophily suggests that actors form positive ties to similar actors and conformity suggests that actors change their opinions to better fit their friends, distancing theory posits that actors form negative ties toward peers that are very different (xenophobia) and then change their opinions to increase cultural differences toward those negative referents. This argument has been addressed by a range of formal modeling studies (Baldassarri and Bearman 2007; Durrett and Levin 2005; Kitts 2006; Macy et al. 2003; Flache and Mäs 2008; Mark 2003), has been studied in extensive experimental research (Berscheid 1966; Sampson and Insko 1964; Schwartz and Ames 1977), and has been applied to social influence through networks in real-world organizations (Kitts 2000).

The second conceptualization of social differentiation postulates that individuals strive for a sufficient feeling of uniqueness (Snyder and Fromkin 1980; Mäs et al. 2010; Imhoff and Erb 2009). Specifically, individuals who feel similar to too many others adjust their opinions and behavior such that they become more distinct, a notion that is also reflected by the theory of "optimal distinctiveness" (Brewer 1991) in social identity research.

While both accounts offer plausible "bottom-up" explanations of social and cultural diversity, they have not yet been systematically compared. To address this lacuna, we present in this paper a formal framework that incorporates both social distancing and striving for uniqueness. We show how both conceptualizations of social differentiation can generate persistent and robust cultural diversity, even within a relatively small and fully-connected network where classical models would predict uniformity. We further demonstrate that the two conceptualizations of differentiation lead to radically different patterns of cultural diversity. Populations of individuals that tend to dislike and thus distance themselves from dissimilar others tend to split into two factions with diametrically opposed opinions, so the entire group is polarized. This is because distancing implies that once sufficiently dissimilar subgroups have formed, members of subgroups strive to increase differences to the members of the other subgroup. Individuals, therefore, tend to develop increasingly extreme opinions. By contrast, striving for uniqueness leads subgroups to seek no more distance from each other than is sufficient to satisfy their desire for uniqueness. Striving for uniqueness creates subgroups with significantly different opinions. However, once these subgroups have formed, opinion differences remain relatively moderate.

Lastly, we show that distancing and striving for uniqueness imply different predictions about the conditions leading to cultural diversity and integration. On the one hand, social distancing increases social diversity only in populations where cultural variation is strong already at the outset of the process. In populations with small initial diversity, individuals perceive few others who are sufficiently dissimilar to generate negative ties and thus motivate distancing. As a consequence, the integrating force of social influence by similar others dominates and opinions move towards consensus. On the contrary, striving for uniqueness is strongest when many individuals hold similar opinions, implying that cultural diversification occurs mainly when there is low cultural diversity.

In closing, we discuss the implications for dynamics of cultural integration in organizations, as well as ways to test the boundary conditions under which the two kinds of social differentiation may shape cultural dynamics in organizations.

5.2 The Model of Social Differentiation

Our agent-based computational model builds on the key assumptions of classical social-influence models (Abelson 1964; Harary 1959; Berger 1981; DeGroot 1974; French 1956; Friedkin and Johnsen 1990; Latane 1981; Lehrer 1975) supplemented with assumptions about social differentiation, conceptualized as either distancing or striving for uniqueness. In the model, each member of the population is represented as an agent i that holds an opinion $o_{i,t}$ which varies continuously between zero and one ($0 \leq o_{i,t} \leq 1$) and can change over time. The social influence and differentiation process is modeled as a sequence of simulation events. At each event t the computer program randomly picks one of the N agents and updates this agent's (i) current opinion $o_{i,t}$ such that after the update a new opinion $o_{i,t+1} = o_{i,t} + \Delta o_{i,t}$ where the magnitude and direction of the opinion change is obtained as

$$\Delta o_{i,t} = \frac{\sum_{j=1}^{N} \left(o_{j,t} - o_{i,t}\right) w_{ij,t}}{\sum_{j=1}^{N} w_{ij,t}} + \xi_{i,t} \,. \tag{5.1}$$

Equation 5.1 includes three key processes that previous models of cultural differentiation have considered: social influence, distancing and striving for uniqueness. The influence weight $w_{ij,t}$ represents the degree to which agent i is influenced by agent j and varies between -1 and $+1$ ($-1 \leq w_{ij,t} \leq 1$). A positive weight implies that j has a positive influence on i, so i's opinion is "pulled" towards the opinion of j. This reflects the mechanism of social influence that has been central to early models of cultural consensus formation (Abelson 1964; DeGroot 1974; French 1956; Lehrer 1975). However, weights can also have negative values, in which case the opinion of agent i is "pushed" away from j's opinion. With negative weights, Eq. 5.1 implements social distancing. Finally, Eq. 5.1 contains a noise term $\xi_{i,t}$ to implement "striving for uniqueness". Specifically, we assume that the less unique an agent's current opinion is in the overall opinion distribution, the larger is the (random) perturbation $\xi_{i,t}$ that leads the agent away from her current opinion. The denominator in (5.1) normalizes influence to ensure that all agents have a fixed capacity to be influenced, apportioned among peers by the tie weights.

Equations 5.2 and 5.3 define the influence weights $w_{ij,t}$. Implementing homophily, we assume that the influence $w_{ij,t}$ that j has on i depends on their opinion distance ($dist_{ij,t} = |o_{i,t} - o_{j,t}|$). To be more precise, Eq. 5.2 implies that the weights are more positive (or less negative) the more similar i and j are. Parameter c ($1 \leq c \leq 2$) allows manipulating the balance of social influence, from positive-only to a mixture of positive and negative influence. If $c = 1$, then influence weights can have only positive values and thus only positive influence operates. If $c = 2$, social distancing (negative influence) is as strong as positive influence. If c

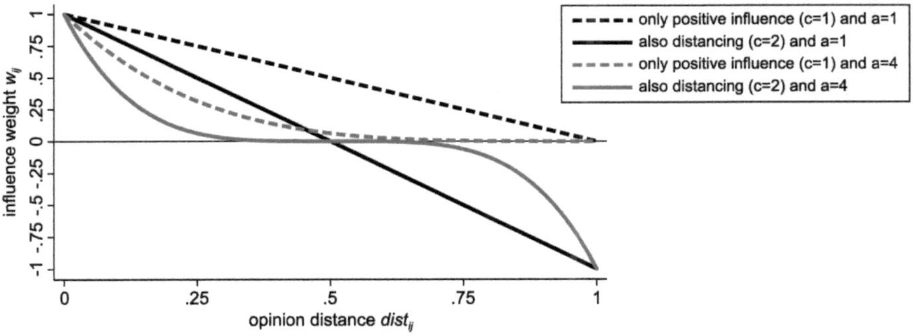

Fig. 5.1 Examples of weight functions for different values of parameters c and a

is between those values then agents are influenced positively ($w_{ij,t} > 0$) by similar others and (to a lesser extent) influenced negatively ($w_{i,j,t} < 0$) by dissimilar others. The value $1/c$ represents the critical opinion distance at which influence shifts from positive to negative.

$$w_{ij} = (1 - c \cdot dist_{ij,t})^a \quad if \quad dist_{ij,t} \leq \frac{1}{c} \tag{5.2}$$

$$w_{ij} = -1(c \cdot dist_{ij,t} - 1)^a \quad if \quad dist_{ij,t} > \frac{1}{c} \tag{5.3}$$

In the case of positive influence ($c = 1$), agents are strongly influenced ($w_{ij,t}$ approaches 1) by peers that are very similar to themselves, and influenced very little ($w_{ij,t}$ approaches 0) by dissimilar agents. When $c = 2$, agents are strongly influenced by very similar peers, strongly negatively influenced ($w_{ij,t}$ approaches -1) by very dissimilar peers, and influenced little ($w_{ij,t}$ approaches 0) by peers that are moderately distant. Parameter a ($a > 0$) allows us to vary the shape of this weight function. In the case of positive influence, high values of a imply that influence diminishes more rapidly with opinion distance, so agents are influenced predominantly by the most similar peers and pay little attention to other peers. In the case of equal positive and negative influence ($c = 2$), high values of a imply that agents are strongly influenced by very similar and also (negatively) by the most dissimilar peers, and pay little attention to the rest.[1] Figure 5.1 illustrates the value of $w_{ij,t}$ resulting from (5.2) and (5.3), under different values of a and c. For illustrative purposes, we have chosen here values of a that are different from those employed in the computational experiments reported further below.

[1]Digital computers may fail to distinguish very small numbers from zero (Izquierdo and Polhill 2006), an error that would be consequential here in that it would erase the distinction between weak influence and no influence. To avoid such problems with floating point inaccuracy, we assign a minimum on positive weights at 10^{-5} and assign a maximum on negative weights at -10^{-5}. We thus conservatively ensure that weak ties are not mistakenly treated as null ties by the computer.

Striving for uniqueness. The second conceptualization of social differentiation assumes that agents adjust their opinions or behavior when they feel indistinguishable from many other individuals. Whereas distancing implies opinion changes away from the opinions of dissimilar others, striving for uniqueness does not specify the direction of the opinion change. Accordingly, we follow the lead of earlier modeling work (Mäs et al. 2010; Pineda et al. 2009) in including noise $\xi_{i,t}$ in updating opinion.

Specifically, a random perturbation is drawn from a normal distribution with an average of zero and a standard deviation specified in (5.4). This implies that striving for uniqueness can result in positive and negative opinion changes with equal probability. Also, small opinion changes tend to be more likely than large changes, incorporating the assumption that greater opinion adjustments imply higher cognitive costs (Festinger 1957; Aronson 1994).

$$\xi_{i,t} = N\left(0, s\sum_{j=1}^{N} e^{-dist_{ij,}}\right) \tag{5.4}$$

Equation (5.4) thus determines the amount of randomness that is added to the agent's opinion, depending on how unique agent i is in the population. If agent i holds an opinion that is very similar to the opinion of many other agents then it feels a stronger need for uniqueness and the standard deviation of the added noise is high. If, however, an agent holds an opinion very different from its peers, it is not driven to increase uniqueness and the standard deviation is low.

We included a parameter $s(s \geq 0)$ that determines the overall degree to which individuals value uniqueness. If $s = 0$, agents do not strive for uniqueness at all. The higher the value of s, however, the stronger is the striving for uniqueness in the population.

Note that distancing may result in opinion values that are outside the defined range of the opinion scale ($0 \leq o_{i,t} \leq 1$). If an agent's opinion would otherwise exceed the range, we assign the extreme value of the range, 0 or 1.

Possible equilibria. Whether model dynamics can reach a state of equilibrium or not and also the number of possible equilibria depends critically on the values assigned to parameters c and s. The model has two possible equilibria if there is only positive or zero influence ($c = 1$) and no striving for uniqueness ($s = 0$). The first equilibrium is characterized by perfect opinion consensus, a state where all agents hold exactly the same opinion.[2] In the second equilibrium, the population consists of two factions of maximally dissimilar extremists. Under this condition,

[2]It is commonly believed that positive influence models invariably produce consensus on connected networks. Even as we add that the network must be strongly connected (i.e. paths allow influence to flow in both directions for all dyads in the population), this may not be strictly true in discrete time for certain network structures if the influence weight is high enough. The lack of convergence is obvious if influence weights ($w_{ij,t}$) are allowed to exceed 1.0, of course, but even $w_{ij,t} = 1$ will yield stable limit cycles that prevent convergence for certain network structures. See Kitts and

opinions cannot change because pairs of agents with nonzero influence hold identical opinions, which implies that opinions remain unaffected. Influence weights between maximally dissimilar agents take the value zero and do not result in opinion changes as well. This replicates the familiar pattern observed in the literature, where uniformity is a strong attractor of the influence dynamic, but distinct subcultures can exist if they are maximally different and have zero influence on one another. We do not further investigate this case here.

If there is distancing ($c > 1$) and no striving for uniqueness ($s = 0$), then multiple equilibria are possible. As with the case of positive influence, global consensus is a locally stable equilibrium; that is, perturbations in the neighborhood of this equilibrium will be self-correcting, and the opinion distribution will return to consensus. Second, this version of the model also implies equilibrium when there are two maximally antagonistic subgroups of extremists. Each extremist is negatively influenced by the agents that adopt the opposite opinion and therefore sticks to the extreme opinion. Unlike in the positive influence case, this polarization equilibrium can be locally stable, and the model will return from small perturbations to the purely polarized state.

Third, the model with distancing and no striving for uniqueness implies that multiplex equilibria can emerge. These equilibria are characterized by opinion distributions with two maximally extreme subgroups and at least one subgroup of moderate agents. In such constellations it is possible that the negative (distancing) and positive influences on the opinions of moderate agents neutralize each other in such a way that agents do not adjust their opinions. For example, assume a population that consists of six agents and is split up into four subgroups, with one agent on each pole of the opinion scale ($o_{1,t} = 0$ and $o_{2,t} = 1$) and two subgroups with moderate opinions ($o_{3,t} = o_{4,t} = 0.375$ and $o_{5,t} = o_{6,t} = 0.625$). Assume furthermore a linear weight function ($a = 1$), and strong distancing ($c = 2$, see Fig. 5.1). In this setting, the two extremists are attracted by two moderate agents. However, this "pull" towards more moderate opinions is overruled by the negative influence of those three agents with very different opinions. As a consequence, the extremists stick to their extreme opinion. Each of the moderate agents is positively influenced by one extremist and negatively influenced by the other. These two influences "pull" the opinion of each moderate towards the nearer extreme. In addition, each moderate agent is positively influenced by the two moderates who belong to the other moderate subgroup. These influences "pull" towards a more moderate opinion with the same strength as the influences of the two extremists, but in the opposite direction. As a consequence, a multiplex equilibrium with more than two co-existing subgroups can arise with distancing, but in the absence of striving for uniqueness.

Trowbridge (2007) for an explanation of the general problem. This is not a danger here because influence weights in our model are strictly determined by similarity; that is, if $w_{ij,t} = 1$ then the agents' opinions are already identical and no influence is possible. Thus stable limit cycles cannot prevent convergence.

If there is striving for uniqueness ($s > 0$), then (5.4) implies that opinions are always exposed to random fluctuations. However, as has been demonstrated by Mäs et al. (2010), the model can reach a dynamic equilibrium, where opinion distributions remain qualitatively similar over a long period of time. Furthermore, if opinion distributions happen to change due to random fluctuations, the system tends to return to a similar state as before the disturbance. Results presented in this paper (cf. Fig. 5.3) further demonstrate this dynamic.

5.3 Results

5.3.1 Ideal-Typical Simulation Scenarios

To begin with, we show a number of typical simulation scenarios that demonstrate the most important differences in the outcomes that the two conceptualizations of social differentiation generate. All of the models presented include a population of 100 agents subject to social influence and homophily, but the differentiation mechanism may be either distancing or striving for uniqueness. Figure 5.2 shows an illustrative simulated trajectory for the model with distancing, but not striving for uniqueness ($c = 2$, $s = 0$). Figure 5.3 shows the model with only striving for

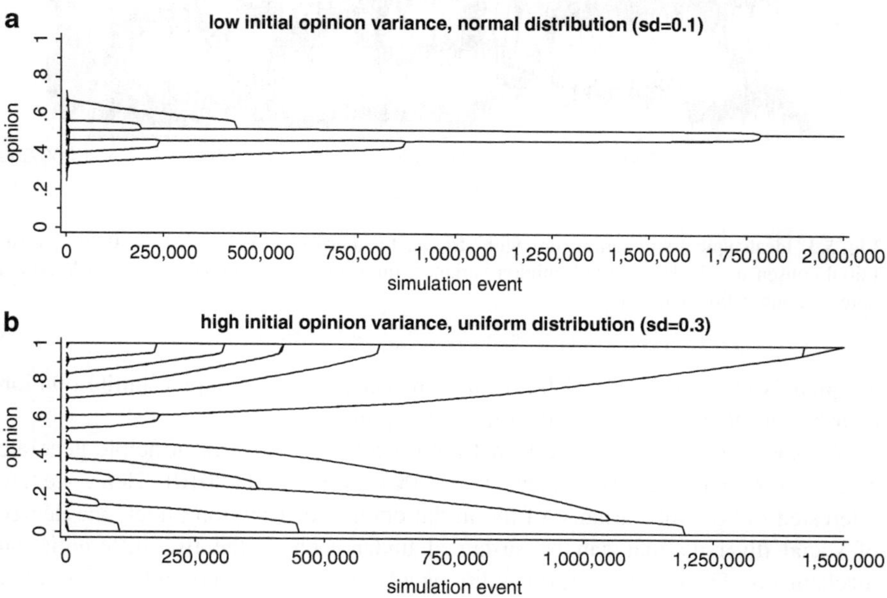

Fig. 5.2 Ideal typical simulation runs with distancing and without striving for uniqueness ($c = 2$, $s = 0$). (**a**) Low initial opinion variance, normal distribution (sd = 0.1). (**b**) High initial opinion variance, uniform distribution (sd = 0.3)

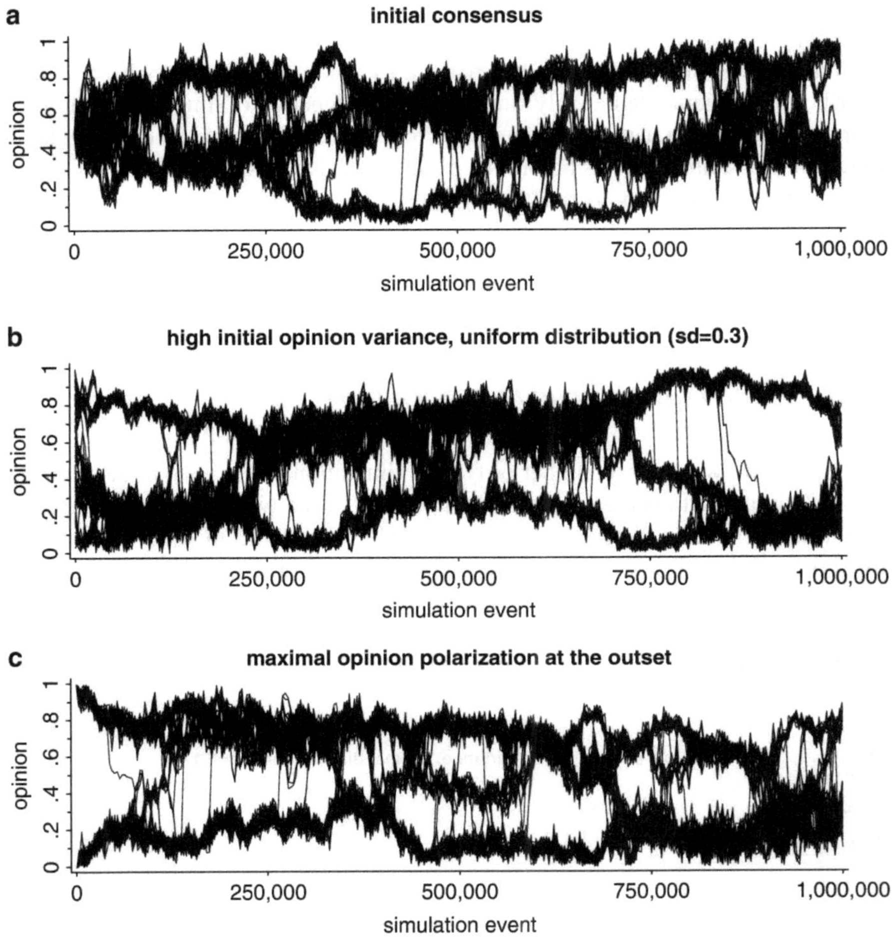

Fig. 5.3 Ideal-typical simulation runs with striving for uniqueness ($c = 1$, $s = 0.00025$). (**a**) Initial consensus. (**b**) High initial opinion variance, uniform distribution (sd = 0.3). (**c**) Maximal opinion polarization at the outset

uniqueness ($c = 1$, $s = 0.00025$). In both scenarios, we compare initial opinion distributions that differ in initial variation in opinions.

We know from previous work that both mechanisms can in principle generate persistent social diversity (Macy et al. 2003; Mäs et al. 2010). Here, we are interested in how the variance of the initial opinion distribution affects the degree of social diversity that can be sustained under each of the two differentiation mechanisms. Therefore, we test conditions where diversity is possible under either mechanism of differentiation. Most importantly, we set a very steep weight function ($a = 100$) because earlier modeling studies (Mäs et al. 2010) demonstrated that this is a critical condition for the formation of distinct subgroups in the uniqueness model. In the uniqueness model, much smaller values result in the formation of a single stable cluster of agents. On the other hand, for much higher a values, the model predicts highly fragmented opinion distributions without any stable cluster formation.

Each panel of Figs. 5.2 and 5.3 shows a line graph where the trajectory of the opinion of each agent is represented by one line. Under both model versions, the social influence process implies that often agents who hold relatively similar opinions from the outset quickly move to identical positions in the opinion space, and then their lines overlap. This is why the initially scattered opinions of agents quickly collapse into a much smaller set of opinions in both models.

Figure 5.2 focuses on the distancing model ($c = 2$) without striving for uniqueness ($s = 0$) and compares influence dynamics that start with a low (panel a) or a high (panel b) initial opinion variance. For the simulation run shown in panel a, we started with a truncated normal opinion distribution with an average of 0.5 and a small standard deviation of 0.1. Accordingly, initial opinion differences in the population were very small, resulting in mainly positive influence weights in the population. Agents with moderate opinions were positively influenced by all others. Only pairs of agents that held opinions near the opposite extremes of the initial opinion distribution had negative influence weights (distancing). However, even these relatively extreme agents were mainly influenced positively by agents with moderate opinions. These positive influences dominated distancing tendencies and the extreme agents then developed moderate opinions. Panel a shows that early in the influence process several subgroups of agents with similar opinions formed. Because of strong homophily, the social influence between agents that belonged to different clusters was weak but eventually led to a steady decrease in opinion differences between subgroups. The model reached a state of equilibrium when all agents converged to the same opinion.

Panel b shows that the outcome of the influence process radically differs if there is initially more opinion variation. For this simulation run, we used the same parameter values as for the run shown in panel a. However, we assumed that the opinion was uniformly distributed in the range (0,1) at the outset, leading more agents to begin with very extreme opinions. In the run shown in panel b, several distinct subgroups formed very early in the influence process, but the extreme agents developed even more extreme opinions over time. This happened because agents with extreme opinions were exposed to influences from multiple agents with very different opinions and tended to distance themselves from those with opposing opinions. Also agents with moderate opinions formed clusters in the early stages of the influence process. Once these subgroups had formed, moderates hardly adjusted opinions because they were exposed to positive influences from agents with both higher and lower opinion values. But as more agents adopted extreme opinions, the moderate agents were also increasingly exposed to negative influences. The figure shows that this resulted in shifts towards extreme opinions also for those who initially maintained moderate positions. Eventually, this process reached equilibrium with two maximally extreme and mutually dissimilar subgroups.

Figure 5.3 depicts three ideal-typical influence scenarios of the model version with striving for uniqueness ($s = 0.00025$) but without distancing ($c = 1$). The scenario shown in panel a started from perfect consensus. Under this condition, positive social influence did not result in opinion adjustments. However, the opinions of the agents were minimally unique. Our implementation of the striving for uniqueness in (5.2) implies ongoing substantial perturbations from the initial consensus. Panel a in Fig. 5.3 shows that these individual opinion perturbations led to a strong increase in overall opinion variation and to the formation of two distinct clusters (e.g. after about 200,000 simulation events).

Once distinct clusters had formed, the composition of each cluster remained temporarily stable. This was because members of each cluster were relatively unique, as there were sufficient opinion differences compared to the members of the other cluster(s). Nevertheless, there were still small individual perturbations from the subgroup consensus, according to (5.4). Because of the strong social influence among the members of an opinion cluster, these small individual perturbations could aggregate to substantial collective opinion changes of all cluster members. It was therefore possible that the members of distinct clusters developed similar opinions and drew the clusters to merge. Once this occurred, the uniqueness of the agents who belonged to the merged subgroup decreased, leading to an increased striving for uniqueness and to the development of new distinct subgroups.

The simulation scenario shown in panel a of Fig. 5.3 demonstrates that the interplay of social influence and striving for uniqueness can create a cyclical fusion and fission of subgroups, which we can call 'factionalism'. In other words, the system tends to develop opinion distributions that consist of several distinct subgroups. However, no distribution is stable because small individual opinion perturbations can lead to fusion of subgroups into larger masses which then break into smaller subgroups again.

Obviously, the differentiation dynamics shown in panel a of Fig. 5.3 differ substantially from those shown in panel b of Fig. 5.2. Most importantly, the distancing mechanism (Fig. 5.2) implies that if dynamics do not end in perfect consensus, the population eventually includes two factions with extreme opinions.[3] However, Fig. 5.3 suggests that the striving for uniqueness mechanism generates clusters with nonextreme opinions.

Another crucial difference between the two conceptualizations of differentiation becomes apparent upon comparing the three simulation scenarios of Fig. 5.3. Panel b shows an ideal-typical simulation scenario that starts out with a uniform opinion distribution. In panel c the initial population consisted of two equally sized and maximally dissimilar subgroups. Even though the three simulation runs shown in Fig. 5.3 started with very different initial opinion distributions, the system always eventually produced the same fusion-and-fission dynamic with similar opinion

[3]We show below that the model may in this condition generate multiplex equilibria, where two extreme factions are accompanied by moderate subgroups. While interesting, these outcomes are very rare and vanish in the presence of noise.

distributions. This apparent robustness to initial conditions contrasts starkly with Fig. 5.2, which demonstrated that the initial distribution of opinions can have a substantial effect on outcomes under the distancing model.

5.3.2 The Computational Experiment

Aim and design of the experiment. The comparison of the ideal-typical simulation scenarios supports that the two conceptualizations of social differentiation imply fundamentally different opinion dynamics. To investigate this conjecture more rigorously, we conducted a computational experiment to see for both versions of the model whether the initial opinion distribution has an impact on the opinion distributions that appear in equilibrium or, for the model with stochastic perturbations, after 25 Million simulation events.

To investigate the effect of the initial opinion distribution, we assumed that the initial opinion of each agent was randomly drawn from a beta distribution. We experimentally manipulated the parameters of the distribution α and β from very low values (yielding low variance in opinions) to very high values (yielding strongly bimodal opinion distributions). Intermediate values of α and β yield approximately normal and approximately uniform distributions of opinion as special cases. We did not include initial distributions where all agents hold exactly the same opinion (perfect consensus) or where the population consists of two maximally distinct subgroups (perfect polarization), considering that these distributions are the equilibria of the model version with negative influence ($c > 1$) and no striving for uniqueness ($s = 0$). In order to generate a sufficient number of experimental conditions with a low, moderate, and high variance, we assigned values to the shape parameters of the beta distribution, α and β, according to Eq. 5.5 and varied the value of m from 1 to 25.

$$\alpha = \beta = 2^{(26-m)/5} - 1 \tag{5.5}$$

Figure 5.4 provides five examples of the opinion distributions that result from this procedure, showing that $m = 1$ results in a very small initial opinion variation with a standard deviation of 0.06 ($\alpha = \beta = 31$). On the opposite extreme, a value of $m = 25$ leads to an almost perfectly polarized opinion distribution with a very high standard deviation of 0.44 ($\alpha = \beta = 0.15$). For all experimental conditions we conducted 100 independent replications. Like in the ideal-typical simulation scenarios, we set $N = 100$ and $a = 100$.

Outcome measures. We used three outcome measures to describe the opinion distributions in the computational experiment. First, we assessed the level of *factionalism* by counting the number of clusters in the distribution of opinions. To identify clusters, we sorted the N agents according to their opinion and defined a

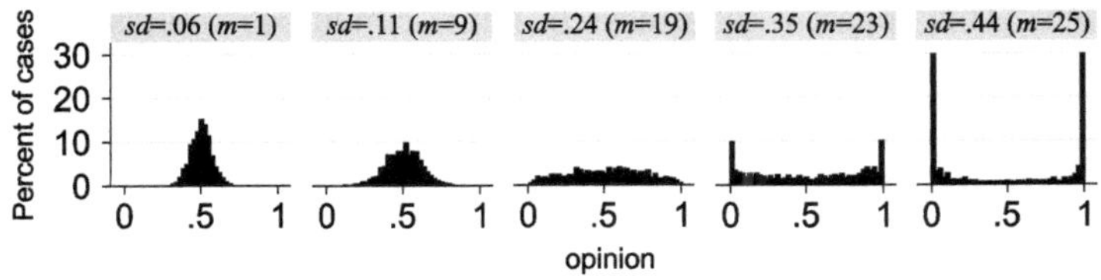

Fig. 5.4 Examples of the initial opinion distributions used in the simulation experiment

subgroup as a set of agents in adjacent positions such that each member of that set was separated from the nearest other member of the set by at most 0.05 scale points. This allows us to identify subgroups of agents with very similar but not identical opinions, which is the appropriate approach for a system in which randomness prevents two agents from having fully identical opinions.

Second, we assessed for each opinion distribution the average *extremeness* in order to test our expectation that differentiation under the distancing mechanism will lead to greater extremity of opinions than differentiation under the uniqueness mechanism. Extremeness was measured as the average distance between an agent's opinion and the mid point of the opinion scale. The resulting value was doubled, normalizing the outcome measure to a scale that ranges from 0 to 1. An average extremeness of 0 indicates that all agents hold an opinion of exactly 0.5. The maximal average extremeness of 1 obtains when all agents hold maximally extreme opinions (0 or 1).

Finally, we used a measure of *polarization* to quantify the degree to which the population splits into mutually distant but internally homogeneous subgroups. Polarization is measured as the standard deviation of the distribution of pairwise opinion distances. Similar to the extremeness measure, we normalized the measure to a scale from 0 to 1. This measure reaches its minimal value of 0 when all agents adopt the same opinion. Its maximal value of 1 obtains if the population is evenly divided into two diametrically opposed subgroups. Thus, polarization implies extreme opinions, but extremeness does not imply polarization.

Results of the computational experiment. Figure 5.5 reports the effect of the standard deviation of the initial opinion distribution on the result of each differentiation process. Circles indicate average values of the outcome measure for the model with distancing. The dashed lines show local polynomial regression lines, describing the relationship between initial opinion variation and the respective outcome measure for the distancing model. Triangles and solid lines report the same statistics for the model with striving for uniqueness.

The solid lines demonstrate that the initial opinion distribution does not have long term effects on the outcome of the differentiation process in the uniqueness model. Panel a shows that the model with striving for uniqueness generated about 2.2 subgroups on average, regardless of the initial opinion distribution. In addition, panel b shows that these subgroups held relatively moderate opinions on average, and panel c shows that opinion polarization is also relatively low.

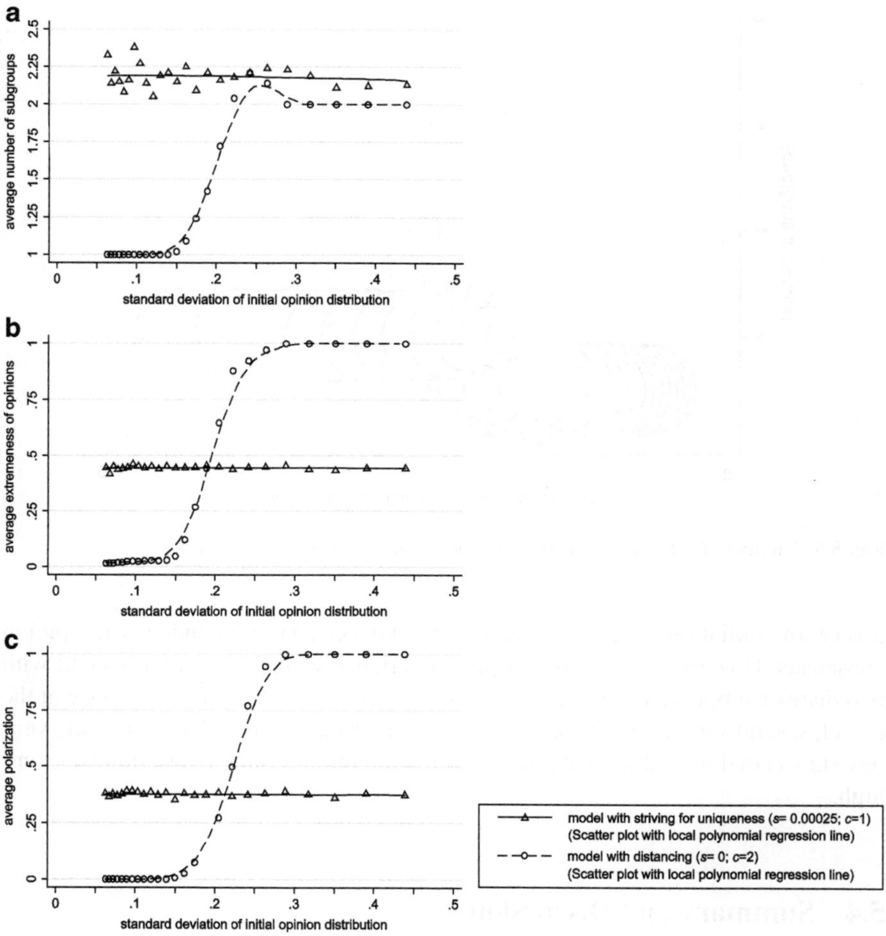

Fig. 5.5 Results of the simulation experiment

The outcome of our experiment is radically different when the differentiation process is driven by distancing (see the dashed lines in Fig. 5.5). Panels a and b show that for initial opinion distributions with a low standard deviation, distancing dynamics tend to generate consensus on moderate opinions. However, higher initial opinion variation resulted in higher extremeness and polarization of opinions. Panel a shows that the average number of subgroups reaches a maximum of about 2.2, indicating that several runs ended in a multiplex equilibrium. To examine this pattern in closer detail, Fig. 5.6 displays the exact distribution of the number of subgroups.

The size of the bubbles in Fig. 5.6 indicates how many simulation runs ended with the respective number of subgroups. In addition, if a bubble represents fewer than 100 runs, the number below or above the bubble reports how many runs ended with the respective number of subgroups. As the figure shows, in conditions with

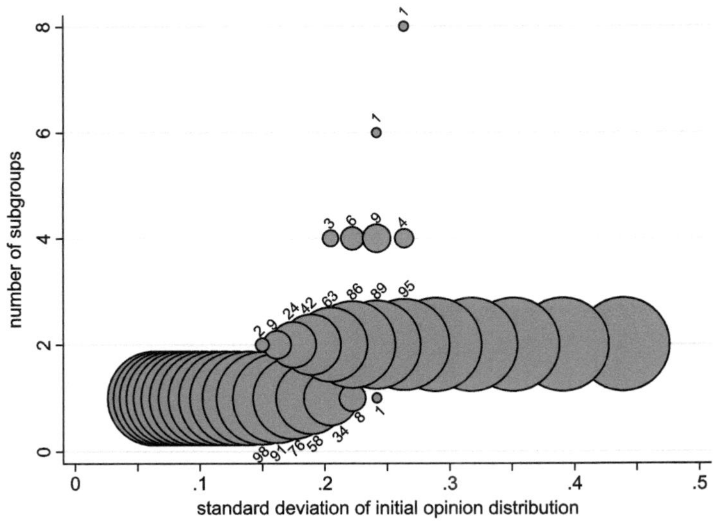

Fig. 5.6 Number of subgroups generated by the distancing model ($c = 2, s = 0$)

a very low initial opinion variation, all 100 runs per condition ended with opinion consensus. However, as the initial opinion variance increases, more runs end with two distinct subgroups. If there was an intermediate level of opinion variance at the outset, several simulation runs ended in a multiplex equilibrium. However, only very few runs ended in multiplex equilibria when the initial opinion variation was very high.

5.4 Summary and Discussion

Classical models of opinion dynamics show that the fundamental mechanism of social-influence – i.e. individuals' tendency to shift their opinions towards those of interaction partners – creates an inexorable march toward cultural homogeneity in connected networks. This contradicts the high degree of persistent diversity that we observe in many social settings, such as in relatively small scale organizations where formal and informal networks are almost guaranteed to be connected. This has led researchers to develop extensions of the classical models to explain emergence and persistence of diversity.

In this contribution, we focused on social differentiation, a recently proposed bottom-up explanation of persistent cultural diversity in strongly connected networks. In particular, we distinguished two alternative conceptualizations of social differentiation – distancing and striving for uniqueness – which operate together with social influence. Distancing implies that individuals tend to form negative ties to others that are very dissimilar, and then differentiate themselves from those negative referents. Striving for uniqueness holds that individuals tend to change their opinions when they perceive that they are not sufficiently different from others. We presented a formal model of social influence dynamics that incorporates both conceptualizations of social differentiation and studied differences in the implications of the two mechanisms.

Our computational experiment demonstrated that these two representations of social differentiation imply radically different patterns of cultural diversity. When individuals distance themselves from dissimilar others, the population may split into two factions with diametrically opposed opinions at the extreme ends of the opinion spectrum. However, striving for uniqueness leads to multiple subgroups with moderate opinions.

In addition, we demonstrated that the two conceptualizations of differentiation imply opposing predictions about the boundary conditions of cultural diversity and integration. On the one hand, distancing increases social diversity only in groups where cultural variation is strong already at the outset of the process. Otherwise, the population approaches uniformity in the long run. On the other hand, striving for uniqueness implies that the degree of cultural diversity in a population is unaffected by the initial distribution.

Both basic processes – distancing and striving for uniqueness – have been independently supported by previous empirical research. It may be that certain individuals are more driven by one force or the other, and it may be that certain situations lead one process or the other to exert a stronger influence. In order to identify different implications of the two conceptualizations of cultural differentiation, we use only a simple stylized model that allows us to examine each of these processes in isolation, and we otherwise hold the situation and the personality of agents constant in our experiments. We recommend that future research should examine both the individual-level and the group-level or situational factors that may moderate the processes that we investigate here.

Of course, distancing and striving for uniqueness may operate interactively in many cases. Our study suggests that this interaction may be quite complex. Remarkably, implications of an integrated model version are very difficult to intuit, as the two differentiation mechanisms have very different implications. For example, distancing implies the development of radicalized subgroups with highly homogeneous opinions and behavior. This, in turn, should motivate individuals who seek to achieve a high level of uniqueness to deviate from their subgroup's consensus and suggests that several individuals who belong to an extreme group will develop more moderate views. However, actors with relatively moderate opinions who are exposed to groups of extremists most likely seek to distance themselves from members of one of the extreme groups and will therefore tend to develop more extreme opinions and values again. Future modeling work is needed to understand the exact implications of the cultural differentiation based on both mechanisms acting in parallel, a research problem that can be tackled based on the formal model which we have presented here.

This paper offers insights into basic processes of cultural influence and differentiation in networks. Although we focus on general, abstract lessons here, a deeper understanding of structural conditions of consensus, clustering, and polarization would be useful to managers or anyone with an interest in how people work together. Empirical research (Jehn 1995) has found that work teams with nonroutine tasks perform relatively poorly when there is no disagreement between team members, suggesting that social differentiation on task-related opinions might be beneficial for work teams as it might trigger inspiring discussions. However, our results suggest that social differentiation in the form of distancing leads to polarized opinions, which has been found to ignite conflicts on work related opinions and hinder team decision making (Jehn 1995, 1997; Jehn and Bendersky 2003). We have demonstrated, on the other hand, that social differentiation based on striving for uniqueness can lead to moderate degrees of diversity. This might create sufficient opinion differences for stimulating discussions and, at the same time, implies enough opinion overlap for efficient team decision-making. Somewhat counter to intuition, this suggests that an organizational culture that supports individuals' striving for uniqueness might actually increase performance of work teams with nonroutine tasks.

Acknowledgements The research of Andreas Flache and Michael Mäs has been supported by the Netherlands Organization for Scientific Research, NWO (VIDI Grant 452-04-351). James Kitts acknowledges support from the National Science Foundation (BCS-0433086 and IIS-0433637).

References

Abelson, R. P.: Mathematical Models of the Distribution of Attitudes Under Controversy. In: Frederiksen, N., Gulliksen, H. (eds.) Contributions to Mathematical Psychology, 142–160. Rinehart Winston, New York (1964)

Aronson, E.: The social animal. Freeman Press, San Francisco (1994)

Axelrod, R.: The dissemination of culture - A model with local convergence and global polarization. J Conflict Res. 41, 203–226 (1997)

Baldassarri, D., Bearman, P.: Dynamics of political polarization. Am Sociol Rev. 72, 784–811 (2007)

Bednar, J., A., Bramson, A., Jones-Rooy, A.,Page, S.: Emergent Cultural Signatures and Persistent Diversity: A Model of Conformity and Consistency. Ration Soc. 22, 407–444 (2010)

Berger, R. L.: A Necessary and Sufficient Condition for Reaching a Consensus Using DeGroot's Method. J Am Stat Assoc. 76, 415–418 (1981)

Berscheid, E.: Opinion Change and Communicator-Communicatee Similarity and Dissimilarity. J Pers Soc Psychol. 4, 670–680 (1966)

Brewer, M. B.: The Social Self - on Being the Same and Different at the Same Time. Pers Soc Psychol B. 17, 475–482 (1991)

Bourdieu, P.: Distinction: A social critique of the Judgment of Taste. Harvard University Press, Cambridge, MA (1984[1979])

Carley, K.: A Theory of Group Stability. Am Sociol Rev. 56, 331–354 (1991)

Carroll, G. R., J. R. Harrison.: Come together? The Organizational Dynamics of Post-Merger Cultural Integration. Simulation Modelling Practice and Theory 10, 349–368 (2002)

Centola, D., Gonzalez-Avella, J. C., Eguiluz, V. M.,Miguel, M. S.: Homophily, cultural drift and the co-evolution of cultural groups. J Conflict Res. 51, 905–929 (2007)

Deffuant, G., Neau, D., Amblard, F.,Weisbuch, G.: Mixing beliefs among interacting agents. Advances in Complex Systems. 3, 87–98 (2000)
DeGroot, M. H.: Reaching a Consensus. J Am Stat Assoc. 69, 118–121 (1974)
De Sanctis, L., Galla, T.: Effects of noise and confidence thresholds in nominal and metric Axelrod dynamics of social influence. Phys Rev E. 79, 046108 (2009)
Durkheim, E.: The Division of Labor in Society. The Free Press, New York (1997 [1893])
Durrett, R., Levin, S. A.: Can Stable Social Groups be Maintained by Homophilous Imitation alone? J Econ Behav Organ. 57, 267–286 (2005)
Elias, N.: The Civilizing Process, Vol.I. The History of Manners. Blackwell, Oxford (1969[1939])
Festinger, L.: A Theory of Cognitive Dissonance. Row, Petersen and Company, Evanston, White Plains (1957)
Festinger, L., Schachter, S., Back, K.: Social Pressures in Informal Groups. Stanford University Press, Stanford, CA (1950)
Flache, A., Macy, M.: Local Convergence and Global Diversity: From Interpersonal to Social Influence. J Conflict Res. 55, 970–995 (2011)
Flache, A., Mäs, M.: How to get the timing right? A computational model of how demographic faultlines undermine team performance and how the right timing of contacts can solve the problem. Comput Math Organ Theory. 14, 23–51 (2008)
French, J. R. P.: A Formal Theory of Social Power. Psychol Rev. 63, 181–194 (1956)
Friedkin, N. E., Johnsen, E. C.: Social-Influence and Opinions. J Math Sociol. 15, 193–205 (1990)
Harary, F.: A criterion for unanimity in French's theory of social power. In: Cartwright, D. (eds.) Studies in social power, 168–182. Institute for Social Research, Ann Arbor (1959)
Harrison, J. R., Carroll, G.: The Dynamics of Cultural Influence Networks. Comput Math Organ Theory. 8, 5–30 (2002)
Harrison, J. R., Carroll, G. R.: Culture and Demography in Organizations. Princeton University Press, Princeton, NJ. (2006)
Hegselmann, R., Krause, U.: Opinion Dynamics and Bounded Confidence Models, Analysis, and Simulation. JASSS-J ARTIF SOC S. 5, (2002)
Heider, F.: Attitudes and Cognitive Organization. In: Fishbein, M. (eds.) Readings in Attitude Theory and Measurement, 39–41. John Wiley and Sons, Inc., New York, London, Sydney (1967)
Homans, G. C.: The Human Group. Harcourt, Brace, and World, New York (1950)
Imhoff, R., Erb, H. P.: What Motivates Nonconformity? Uniqueness Seeking Blocks Majority Influence. Pers Soc Psychol B. 35, 309–320 (2009)
Izquierdo, L. R., Polhill, J. G.: Is Your Model Susceptible to Floating-Point Errors? JASSS-J ARTIF SOC S. 9 (2006)
Jehn, K. A.: A Multimethod Examination of the Benefits and Detriments of Intragroup Conflict. Admin Sci Quart. 40, 256–282 (1995)
Jehn, K. A.: A Qualitative Analysis of Conflict Types and Dimensions of Organizational Groups. Admin Sci Quart. 42, 530–557 (1997)
Jehn, K. A., Bendersky, C.: Intragroup conflict in organizations: A contingency perspective on the conflict-outcome relationship. Res Organ Behav. 25, 187–242 (2003)
Kitts, J. A.: Mobilizing in Black Boxes: Social Networks and Participation in Social Movement Organizations. Mobilization. 5, 241–257 (2000)
Kitts, J.: Social Influence and the Emergence of Norms Amid Ties of Amity and Enmity. Simulation Modelling Practice and Theory. 14, 407–422 (2006)
Kitts, J. A., Trowbridge, P. T.: Shape Up Or Ship Out: Social Networks, Social Influence, and Organizational Demography. Comput Math Organ Theory. 13, 333–353 (2007)
Klemm, K., Eguiluz, V. M., Toral, R.,Miguel, M. S.: Global culture: A noise-induced transition in finite systems. Phys Rev E. 67, 045101(R) (2003)
Lehrer, K.: Social Consensus and rational agnoiology. Synthese. 31, 141–160 (1975)
Latane, B.: The Psychology of Social Impact. Am Psychol. 36, 343–356 (1981)

Latane, B.: Pressures to uniformity and the evolution of cultural norms. In: Ilgen, D. R., Hulin, C. L. (eds.) Computational Modeling of Behavior in Organizations: The Third Scientific Discipline, 189–215. American Psychological Association, Washington, DC (2000)
Macy, M. W., Kitts, J., Flache, A.,Benard, S.: Polarization and Dynamic Networks. A Hopfield Model of Emergent Structure. In: Breiger, R., Carley, K., Pattison, P. (eds.) Dynamic Social Network Modeling and Analysis: Workshop Summary and Papers, 162–173. The National Academies Press, Washington, DC (2003)
Mark, N. P.: Beyond individual differences: Social differentiation from first principles. Am Sociol Rev. 63, 309–330 (1998)
Mark, N. P.: Culture and Competition: Homophily and Distancing Explanations for Cultural Niches. Am Sociol Rev. 68, 319–345 (2003)
Mäs, M., Flache, A.,Helbing, D.: Individualization as Driving Force of Clustering Phenomena in Humans. PLoS Computational Biology. 6, e1000959 (2010)
Parisi, D., Cecconi, F.: Cultural Change in Spatial Environments. J Conflict Res. 47, 163–179 (2003)
Pineda, M., Toral, R.,Hernandez-Garc?a, E.: Noisy continuous-opinion dynamics. Journal of Statistical Mechanics. P08001 (2009)
Sampson, E. E., Insko, C. A.: Cognitive consistency and performance in the autokinetic situation. Journal of Abnormal and Social Psychology. 68, 184–192 (1964)
Schwartz, S. H., Ames, R. E.: Positive and Negative Referent Others as Sources of Influence: A Case of Helping. Sociometry. 40, 12–21 (1977)
Shibanai, Y., Yasuno, S.,Ishiguro, I.: Effects of Global Information Feedback Diversity. J Conflict Res. 45, 80–96 (2001)
Snyder, C. R., Fromkin, H. L.: Uniqueness. The human Pursuit of Difference. Plenum Press, New York and London (1980)
Tajfel, H., Turner, J. C.: The Social Identity Theory of Intergroup Behavior. In: Worchel, S., Austin, W. G. (eds.) Psychology of Intergroup Relations, 7–24. Nelson-Hall Publishers, Chicago (1986)

Chapter 6
Modeling and Analysis of Safety Culture of Air Traffic Organizations in the National Culture Context

Alexei Sharpanskykh and Sybert H. Stroeve

6.1 Introduction

Safety culture is broadly recognized as important for operational safety in various fields, including air traffic management, power plant control and healthcare. Currently, as a prelude to systemic changes in air traffic management via new programs SESAR in Europe and NextGen in the USA more and more Air Navigation Service Providers (ANSPs) go through safety culture improvement processes.

There exists a variety of definitions of safety culture. We use the term *safety culture* as those aspects of organizational culture that may have an effect on safety, in line with reasoning of Hopkins (2006). Following the definition by Uttal (1983), among those aspects are safety-related shared values (what is important) and beliefs (how things work) that interact with a company's people, organizational structures and control systems to produce safety-related behavioral norms (the way we do things around here). It is widely recognized that the type of national culture has a significant impact on organizational and safety cultures (Choudhry et al. 2007; Ek et al. 2007; Hopkins 2006). However, identifying specific types of influences of national culture characteristics on organizational structures and processes is not trivial. To address this challenge we propose a formal, agent-based approach for modeling organizations (in particular, ANSPs), which establishes explicit relations between safety culture, national culture and organizational processes and structures. The approach incorporates national cultural characteristics from the cultural classification framework by Hofstede (2005), and identifies their direct and

A. Sharpanskykh (✉)
Vrije Universiteit Amsterdam, De Boelelaan 1081a, Amsterdam, The Netherlands
e-mail: sharp@few.vu.nl

S.H. Stroeve
National Aerospace Laboratory NLR, Anthony Fokkerweg 2, 1059 CM Amsterdam, The Netherlands
e-mail: stroeve@nlr.nl

V. Dignum and F. Dignum (eds.), *Perspectives on Culture and Agent-based Simulations*, Studies in the Philosophy of Sociality 3, DOI 10.1007/978-3-319-01952-9_6,

indirect influences on individual and organizational behavior. In comparison with other existing cultural frameworks (e.g., by Trompenaars 1997), the framework by Hofstede (2005) is more suitable to characterize professional relations existing in air traffic organizations. Furthermore, the Hofstede's dimensions were reflected well in empirical data provided by such organizations for our study.

The approach was applied in a case study on safety occurrence reporting in an ANSP, which is considered in this chapter. Air traffic controllers in an ANSP are obliged to report safety occurrences observed during air and ground operations. Knowledge about safety occurrences is useful for identification and management of safety problems in ANSPs. In practice, however, safety occurrences are not always reported, which may create a serious bottleneck in the organizational safety. It is recognized that there is a strong reciprocal relation between the organizational safety culture and the reporting behavior of air traffic controllers (Ek et al. 2007). In the models developed in the study this relation is elaborated formally, in detail.

Based on domain knowledge obtained from existing ANSPs (organizational documents, interviews, and domain literature), a prototypical organizational model of safety occurrence reporting in an ANSP was built (Sharpanskykh and Stroeve 2011). This generic model was instantiated to create simulation models of Eastern and Western European ANSPs of different types, discussed in this chapter. A model for an existing Western European ANSP developed in this study was validated successfully (Stroeve and Sharpanskykh 2009; Stroeve et al. 2011). Based on the developed models, the safety culture of several types of air navigation service provider organizations from Western and Eastern European cultures was investigated by simulation. Results of this comparative simulation study are described in the chapter.

The chapter is organized as follows. A safety occurrence reporting case and related safety culture indicators are introduced in Sect. 6.2. A generic model of the formal organization for this case is considered in Sect. 6.3. A specification of the behavior of organizational agents is described in Sect. 6.4. Relations between the safety culture indicators and the developed model are described in Sect. 6.5. Simulation results for models of ANSPs instantiated in the Western and Eastern cultural contexts are presented in Sect. 6.6. Section 6.7 concludes the chapter.

6.2 Case Study

In operations performed by air traffic controllers hazardous situations may occur, such as separation minima infringements and runway incursions. In the context of this study we distinguish four types of occurrences:

- Type A: incident with serious inability to provide or maintain safe service, involving a large separation infringement and a high risk of a collision;
- Type B: incident with partial inability to provide or maintain safe service, involving a medium separation infringement and a medium risk of collision;

- Type C: incident with ability to provide or maintain safe service, involving a small separation infringement and a low risk of collision;
- Other: occurrence without separation infringement, e.g. pilot report of a traffic collision avoidance system advisory or a prolonged loss of communication.

According to air traffic safety regulations, controllers are obliged to report safety occurrences of a large number of predefined types. Information about safety occurrences is useful for safety analysis, such as identification of safety trends. Some ANSPs provide reprimands to controllers for occurrences. Sometimes also rewards are provided for reporting occurrences. Although reporting is obligatory, not all identified occurrences may be reported properly by controllers. To understand the reasons for such behavior, the occurrence reporting cycle, including the controller's decision making whether to report an occurrence, is modeled in this study.

Controllers work in shifts, within each shift a pair of controllers is allocated to each air traffic control sector. A controller supervisor manages work in a shift. After a controller decides to report an observed occurrence, s/he creates a notification report. The draft report is reviewed and maybe corrected by the supervisor and it is provided to the Safety Investigation Unit of the ANSP. Depending on the occurrence severity and the collected information about (similar) occurrences, the Safety Investigation Unit makes the decision whether to initiate a detailed investigation. During the investigation accumulated organizational knowledge and data about safety related issues (in particular, learned from notification reports) is used. As the investigation result, a final occurrence assessment report is produced, which should be provided to the controller-reporter as a feedback. Furthermore, often final reports contain recommendations for safety improvement, which should be implemented by an ANSP (e.g., provision of training, improvement of formal procedures, extension of staff).

To evaluate the ANSP's safety culture in relation to safety occurrence reporting quantitatively, a set of safety culture indicators was identified (Sharpanskykh and Stroeve 2008):

SCI1: Average reporting quality of controllers. It refers to the ratio of reported to observed occurrences.

SCI2: Average quality of the processed notification reports. It refers to the correctness and completeness of information about the reported occurrences.

SCI3: Average quality of the final safety occurrence assessment reports. It refers to the completeness of the occurrence report with respect to the causes of the occurrence.

SCI4: Average quality of the monthly safety overview reports received by controllers. It refers to the completeness of the report with respect to the safety trends.

SCI5: Average commitment to safety of controllers.

SCI6: Average commitment to safety of management as perceived by controllers.

6.3 Modeling the Formal Organization

A model for the formal organization is built along the three organizational views (Sharpanskykh 2008; Sharpanskykh and Stroeve 2011): *a process-oriented view* describes organizational workflows as well as static structures of tasks and resources; *a performance-oriented view* is characterized by a goal structure, a performance indicators structure, and relations between them as well as relations between goals and tasks, performance indicators and processes, goals and roles or agents; *an organization-oriented view* defines organizational roles, each associated with a set of tasks and characterized by authority and responsibility relations on tasks, resources and information. Commitment, obligation and power relations and sets of competences required for agent allocation to roles are also defined. To express structural organizational relations sorted predicate logic-based languages are used, whereas the Temporal Trace Language (TTL) (Sharpanskykh and Treur 2010) is used for specifying dynamic aspects of organizations. In the following a generic model for the formal organization from the case study is partially presented. A complete description of the model is provided in (Sharpanskykh et al. 2008; Sharpanskykh and Stroeve 2011).

The Identification of the Organizational Roles A *role* is a (sub-)set of functionalities of an organization, which are abstracted from specific agents who fulfill them. Each role can be composed by several other roles, until the necessary detailed level of aggregation is achieved. A formal or informal organizational group is modeled as a composite role. At a higher abstraction level, sets of simple roles clustered in a composite role act as a single entity interacting with other (composite) roles. Each role has an input and an output interface, which facilitate interaction with other roles. The environment represents a special component of a model, which also has input and output interfaces. In the case study roles are identified at three aggregation levels. At the aggregation level 1 the Air Navigation Service Provider is considered as one composite role. The subroles of the Air Navigation Service Provider are described at the aggregation level 2 (see Fig. 6.1).

The Specification of the Interactions Between the Roles Relations between roles are represented by interaction and interlevel links. An *interaction link* is an information channel between two roles at the same aggregation level. An *interlevel link* connects a composite role with one of its subroles. The interaction relations at the ANSP's aggregation levels 2 is shown in Fig. 6.1.

The Identification of the Requirements for the Roles In this step the requirements on knowledge, skills and personal traits of an agent implementing a role are identified. A prerequisite for the allocation of an agent to a role is the existence of a mapping between the capabilities and traits of the agent and the role requirements. For example, among the requirements for the controller role are: passed a rigid medical examination; thorough knowledge of the flight regulations; air traffic control training; excellent communication skills.

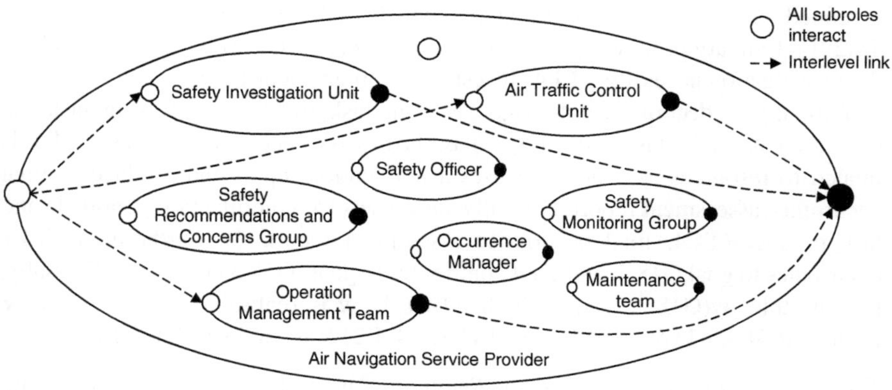

Fig. 6.1 The interaction relations between the subroles of the role Air Navigation Service Provider at the aggregation level 2

The Identification of the Organizational Performance Indicators and Goals A performance indicator (PI) is a quantitative or qualitative indicator that reflects the state/progress of the company, unit or individual. PIs can be hard (e.g., occurrence investigation time) or soft, i.e., not directly measurable, qualitative (e.g., level of collaboration between controllers). *Goals* are objectives that describe a desired state or development and are defined formally as expressions over PIs. A goal can be refined into subgoals forming a hierarchy. For example, goal G18 'It is required to maintain timeliness and a high quality of occurrence investigation' is based on two PIs 'timeliness of occurrence investigation' and 'quality of occurrence investigation'. This goal is refined in several subgoals: G18.1 'It is required to maintain a high proficiency level of incident investigators', G18.2 'It is required to maintain a sufficient level of details of notification reports', G18.3 'It is required to maintain the timely investigation of an occurrence' and G18.4 'It is required to maintain a high level of thoroughness of occurrence investigation'. Goals are related to roles. For example, G18 is attributed to Safety Investigation Unit role of the ANSP.

The Specification of the Resources Resource types are characterized by: *name*, *category*: discrete or continuous, *measurement unit*, *expiration duration*: the time interval during which a resource type can be used; *location*; *sharing*: some processes may share resources. Examples of resource types are: aircraft, incident classification database.

The Identification of the Tasks and Relations Between the Tasks, the Resources and the Goals A task represents a function performed in the organization and is characterized by *name*, *maximal* and *minimal duration*. Tasks can be decomposed into more specific ones using AND- and OR-relations. Each task performed in an organization should contribute to the satisfaction of one or more organizational goals. For example, task T4 'Safety occurrence reporting and the report handling' is refined into more specific tasks, among which T4.1 'Create a notification report', T4.2 'Preliminary processing of a notification report', T4.3 'Making decision about the investigation necessity', T4.4 'Investigation of an occurrence'.

Formally, is_decomposed_to(T4, L) & is_in_task_list(T4.1, L) & is_in_task_list (T4.2, L) & is_in_task_list(T4.3, L) & is_in_task_list(T4.4, L). Task T4.4 is related to resources as follows: it uses a notification report and produces a final occurrence assessment report. Formally, task_uses(T4.4, notification_report, 1) and task_produces(T4.4, final_occurrence_assessment_report, 1). Furthermore, T4.1 contributes to goal G18.2, and T4.4 contributes to goals G18.3 and G18.4. Formally, is_realisable_by(G18.2, L) & is_in_list(T4.1, L) & is_realisable_by(G18.3, L1) & is_in_list(T4.4, L1) & is_realisable_by(G18.4, L2) & is_in_list(T4.4, L2).

The Specification of the Authority Relations The following types of authority relations are distinguished: superior-subordinate relations on roles with respect to tasks, responsibility relations, control for resources, authorization relations. Roles may have different rights and responsibilities with respect to different aspects of task execution, such as execution, passive monitoring, consulting, making technological and managerial decisions. For example, Safety Investigator role is assigned responsible for execution of and making technological decisions with respect to task T4.4, Safety Manager is responsible for monitoring, consulting and making managerial decisions related to T4.4. Formally, is_authorized_for(Safety Investigator, execution, T4.4) & is_authorized_for(Safety Investigator, tech_decisions, T4.4) & is_authorized_for(Safety Manager, monitoring, T4.4) & is_authorized_for(Safety Manager, consulting, T4.4) & is_authorized_for(Safety Manager, manag_decisions, T4.4).

The Specification of the Workflows Workflows describe temporal ordering of processes of an organization in particular scenarios. A workflow starts with the process BEGIN and ends with the process END; both have zero duration. The (partial) order of execution of processes in the workflow is defined by sequencing, branching, cycle and synchronization relations (referred to as ordering relations) specified by the language from (Popova and Sharpanskykh 2008).

This workflow is executed every time when an occurrence has been identified by a controller. In the following the workflow is considered briefly. After a controller decides to report an observed occurrence, she creates a notification report, which is provided to the Safety Investigation Unit (SIU). Depending on the occurrence severity and the collected information about similar occurrences, Safety Investigator role in SIU makes the decision whether to initiate a detailed investigation. During the investigation, accumulated organizational data and knowledge about safety related issues is used. As the investigation result, a final occurrence assessment report is produced, which provides feedback to the controller-reporter. Furthermore, often final reports contain recommendations for safety improvement, which are advised to be implemented by the ANSP.

This workflow is executed continuously in a cycle, i.e., whenever the current instance of the workflow finishes, a new instance of the workflow starts.

Table 6.1 Organizational reprimand policies used in simulation

Low personal consequences	repr(1, A) = 1
Medium personal consequences	repr(1, A) = 1; repr(1, B) = 0.5
High personal consequences	repr(1, A) = 1; repr(1, B) = 0.5; repr(2, C) = 0.2; repr(4, other) = 0.1

The Identification of the Generic and Domain-Specific Constraints Generic constraints ensure internal consistency of an organizational specification. Domain specific constraints restrain behavior of individuals in a particular organization. In particular, organizational reward policies can be formalized as domain-specific constraints. For example, the reprimand policy for reporting can be formalized by a set of constraints using a function repr that maps the number of occurrences of some type to a reprimand value [0, 1]: repr(1, A) = 1; repr(1, B) = 0.5.

Table 6.1 lists three reprimand policies with the increasing levels of personal consequences used in simulation. For example, if a controller was involved in 1 occurrence of type B and 2 occurrences of type C during an evaluation period in an ANSP in which the reprimand policy with high personal consequences is applied, then the controller will receive the reprimand value 0.7.

6.4 Modeling Organizational Agents in a National Culture

The specification of a formal organization forms a part of an overall organizational description. Another part describes the characteristics and behavior of agents.

The behavior of agents is considered from external and internal perspectives, which both are described in this section.

From the external perspective the behavior is specified by temporal correlations between agent's input and output states, corresponding to interaction with other agents and with the environment. Agents perceive information by observation and generate output in the form of communication or actions. Since agents are allocated to organizational roles, communication among them is specified using the interaction ontologies of roles. Furthermore, an agent observes the behavior of other agents and of the environment. To represent communication between humans, speech act theory is used. This theory allows representing a wide diversity of illocutionary speech acts constituting communication.

In the following the internals of an agent are considered. It is widely recognized in the literature (Pinder 1998) that the behavior and dynamics of internal processes of a human are influenced by *personal traits* and *capabilities*. In general, one can specify a great variety of personal traits (e.g., the big five model Goldberg 1993; Carver and Scheier 2000). In particular, cultural traits are often recognized as influential for human behavior (Hofstede 2005). In the case study, cultural traits are specified for each air traffic controller based on the cultural classification framework by (Hofstede 2005) (see Table 6.2). More specifically, the following indexes from

Table 6.2 The ranges of the uniformly distributed cultural indexes for controllers from the Western and Eastern European cultures used in simulation

Agent type	IDV	PDI	UAI
Eastern culture	[0.2, 0.4]	[0.8, 1]	[0.8, 1]
Western culture	[0.7, 0.9]	[0.3, 0.5]	[0.4, 0.6]

Table 6.3 The minimal acceptable satisfaction values of needs and the ranges of the uniformly distributed basic valences of needs of controller agents used in the simulation

Culture	min_accept (n1)	min_accept (n2)	min_accept (n3)	min_accept (n4)	min_accept (n5)
Eastern culture	1	0.8	1	0.7	0.6
Western culture	1	0.6	0.5	0.7	0.9

the framework are used: *individualism* (IDV) is the degree to which individuals are integrated into groups; *power distance index* (PDI) is the extent to which the less powerful members of an organization accept and expect that power is distributed unequally; and *uncertainty avoidance index* (UAI) deals with individual's tolerance for uncertainty and ambiguity. The domain of each index is [0,1].

Interviews indicated that the macho-culture becomes weaker in the community of air traffic controllers now, therefore the cultural dimension *masculinity* is less relevant for our study. Furthermore, the distinction in *long-term orientation* (LTO) is primarily of importance for studying differences between European and Eastern Asian countries. This index is almost the same for the European cultures considered in our study. Therefore, it was not considered in the model.

Human capabilities include *knowledge* and *skills*. Knowledge of a human is a set of reasoning procedures, which are used for execution of organizational tasks. *Skills* describe developed abilities of a human to use effectively and readily his or her knowledge for the performance of tasks.

As generally assumed in social science the behavior of agents is considered to be goal-directed. Furthermore, the goals of an agent are based on its needs. Different types of needs are distinguished (Pinder 1998): (n1) extrinsic needs associated with biological comfort and material rewards; social interaction needs that refer to the desire for social approval and affiliation, in particular (n2) own group approval, and (n3) management approval; intrinsic needs that concern (n4) the desires for self-development and self-actualization, in particular contribution to organizational safety-related goals and (n5) self-confidence and self-actualization needs. Different needs have different priorities and minimal acceptable satisfaction levels for individuals in different cultures. The minimal acceptable satisfaction values of the needs of controller agents used in the simulation provided in Table 6.3 were defined in correlation with the individual characteristics from Table 6.2: for n2 – with IDV, for n3 – with PDI and UAI, for n4 – with UAI.

It is assumed that the agents in both cultures have the highest minimal acceptable satisfaction value (1) for the extrinsic needs. In line with (Pinder 1998; Hofstede

2005), the realization of self-actualization needs in the Eastern culture is less important than own group approval. On the other hand, social approval is less important than self-actualization and self-development in the individualistically-oriented Western culture. This is reflected in the minimal acceptable satisfaction values for needs n3 and n5 in Table 6.3.

Needs exert a significant influence on the cognitive dynamics of an agent, in particular on its decision making.

From the internal (cognitive) perspective the behavior is characterized by a specification of direct causal relations between internal states of the agent, based on which an externally observable behavioral pattern is generated. Such types of specification are called causal networks. The relations in the causal networks are formalized by temporal equations. Some of these equations are instantaneous, whereas others are difference equations over successive time steps. In such a way one can specify mutual influences of states and loops in cognitive processes of agents. Based on the air traffic domain literature and social science literature a number of causal networks have been identified, in particular for cognitive states as commitment to safety, maturity w.r.t. organizational tasks and attitude to reporting. All of these cognitive structures are described in detail in (Sharpanskykh et al. 2008). In the following different types of internal states of agents are considered that form such causal networks, used further in decision making.

It is assumed that agents create time-labeled internal representations (*beliefs*) about their input and output states, which may persist over time:

$$\forall \mathrm{ag}: \mathrm{AGENT}\ \forall \mathrm{p}: \mathrm{STATE\_PROPERTY}\ \forall \mathrm{t}: \mathrm{TIME}\ \mathrm{at}\,(\mathrm{input}\,(\mathrm{ag}, \mathrm{p}), \mathrm{t}) \rightarrow$$
$$\mathrm{at}\,(\mathrm{internal}\,(\mathrm{ag}, \mathrm{belief}\,(\mathrm{p}, \mathrm{t}), \mathrm{t}+1))$$

Information about observed safety occurrences is stored by agents as beliefs: e.g., belief(observed_occurrence_with(ot: OCCURRENCE_TYPE, ag:AGENT)), t:TIME).

Besides beliefs about single states, an agent forms beliefs about dependencies between its own states, observed states of the environment, and observed states of other agents:

belief(occurs_after(p1:STATE_PROPERTY, p2:STATE_PROPERTY, t1:TIME, t2:TIME), t:TIME), which expresses that state property p2 holds t' ($t1 < t' < t2$) time points after p1 holds.

In the considered case each controller agent creates the belief about the dependency between providing of a notification report on an occurrence of some type to his/her supervisor and receiving a final assessment report on the occurrence (i.e., feedback) from a safety investigator agent. Moreover, often final assessment reports include recommendations for organizational and environmental improvement, which when implemented may be observed by the controller-reporter. An agent may have beliefs about faulty dependencies and/or make incorrect estimation of likelihoods (e.g., some controllers may believe that reporting of insignificant occurrences is punishable, whereas in reality it is not).

Another internal state highly relevant for decision making is the agent's *commitment to safety*. In the following a causal network that forms this state is discussed. Commitment to safety is determined largely by the agent's maturity degree w.r.t. the agent's tasks (Hersey et al. 2001). In the theory of situational leadership (Hersey et al. 2001) the agent's maturity w.r.t. to a task is defined as an aggregate of the agent's experience, willingness and ability to take responsibility for the task. The agent's willingness to perform a task is determined by the agent's confidence and commitment, which are necessary for the ATC task execution. The ability of an agent to perform a task is determined by its knowledge and skills. Thus, the agent's maturity is a complex notion, which value is calculated based on other variables of the model. Furthermore, the maturity value changes over time as a result of gaining new knowledge and skills, and changing self-confidence of a controller. In an efficient, committed to safety ANSP the maturity of a controller grows until some high value is reached and then fluctuates slightly around this value.

In the model, the adequacy of the mental models for the air traffic control (ATC) tasks depends on the sufficiency and timeliness of training provided to the controller and the adequacy of knowledge about safety-related issues. Such knowledge is contained in reports that resulted from safety-related activities: final occurrence assessment reports resulted from occurrence investigations and monthly safety overview reports.

Many factors influence the quality of such reports, for specific details we refer to (Sharpanskykh et al. 2008). Thus, the maturity level of a controller agent ($e5_{a,\,t}$) is calculated as:

$$e5_{a,t} = w22 \cdot e19_{a,t-1} + w23 \cdot e20_{a,t-1} + w24 \cdot e21_{a,t-1} + w25 \cdot e23_{a,t-1}$$

here $e19_{a,t}$ is the agent's self-confidence w.r.t. the ATC task (depends on the number of occurrences with the controller); $e20_{a,t}$ is the agent's commitment to perform the ATC task; $e21_{a,t}$ is the agent's development level of skills for the ATC task; $e23_{a,t-1}$ is the adequacy of the mental models for the air traffic control (ATC) tasks.

The agent's commitment to safety is also influenced by the perceived commitment to safety of other team members and the management. An agent evaluates the management's commitment to safety by considering factors that reflect the management's effort in contribution to safety (investment in personnel and technical systems, training, safety arrangements).

In such a way, the commitment value is calculated based on a feedback loop: the agent's commitment influences the team commitment, but also the commitment of the team members and of the management influence the agent's commitment:

$$e6_{a,t} = w1 \cdot e1_{t-1} + w2 \cdot e2_{a,t-1} + w3 \cdot e3_{a,G,t-1} + w4 \cdot e4_{a,t-1} + w5 \cdot e5_{a,t-1},$$

here $e1_t$ is the priority of safety-related goals in the controller's role description, $e2_{a,t}$ is the perception of the commitment to safety of management, $e3_{a,G,t}$ is the perception of the average commitment to safety of the team, $e4_{a,t}$ is the perceived influence degree of controller a on safety arrangements, $e5_{a,t}$ is the controller's maturity level w.r.t. the task; w1–w5 are the weights.

Table 6.4 The ranges of the uniformly distributed cultural indexes for controllers from the Western and Eastern European cultures used in simulation.

Agent type	w1	w2	w3	w4	w5
Eastern culture	[0.1, 0.3]	[0.2, 0.4]	[0.2, 0.4]	[0, 0.2]	[0, 0.2]
Western culture	[0.05, 0.2]	[0.1, 0.3]	[0.05, 0.2]	[0.2, 0.4]	[0.2, 0.4]

Based on the literature and domain knowledge, some of the weights of influence relations used in these causal networks were defined as dependent on the national culture. In particular, the weights w1–w5 used for $e6_{a,t}$ depend on the national culture, as shown in Table 6.4. Specifically, the ranges for w1 reflect a greater influence of the formal organization on a controller in the Eastern culture in comparison with the Western culture; the ranges for w2 reflect a higher PDI in the Eastern culture in comparison with the Western culture; the ranges for w3 reflect a lower IDV for the Eastern culture in comparison with the Western culture (i.e., a higher team influence); the ranges for w4 reflect the domain knowledge that the perceived influence degree of a controller on safety arrangements is more important for the controller's commitment to safety in Western European cultures, than in Eastern European cultures; the ranges for w5 reflect a higher IDV for the Western culture in comparison with the Eastern culture.

Cognitive states from the identified causal network are used in decision making. To model agent decision making a refined version of the expectancy theory by Vroom (Pinder 1998) has been used. Some advantages of the expectancy theory are: (a) it can be formalized; (b) it allows incorporating the organizational context; (c) it has received good empirical support (Pinder 1998). According to this theory, when a human evaluates alternative possibilities to act, s/he explicitly or implicitly makes estimations for the following factors: valence, expectancy and instrumentality. In Fig. 6.2 the decision making model for reporting a safety occurrence is shown.

Expectancy refers to the individual's belief about the likelihood that a particular act will be followed by a particular outcome (called a first-level outcome). For example, E{1,2} in Fig. 6.2 refers to the agent's belief of how likely that reporting of an occurrence will be followed by an administrative reprimand. Instrumentality is a belief concerning the likelihood of a first level outcome resulting into a particular second level outcome; its value varies between -1 and $+1$. Instrumentality takes negative values when a second-level outcome does not follow a first-level outcome. A second level outcome represents a desired (or avoided) state of affairs that is reflected in the agent's needs. For example, I{1,3,2} in Fig. 6.2 refers to the belief about the likelihood that own group appreciation of the action results in own group approval. In the proposed approach the original expectancy model is refined by considering specific types of individual needs, described above. Valence refers to the strength of the individual's desire for an outcome or state of affairs; it is also an indication of the priority of needs. Values of expectancies, instrumentalities and valences change over time, in particular due to individual and organizational learning.

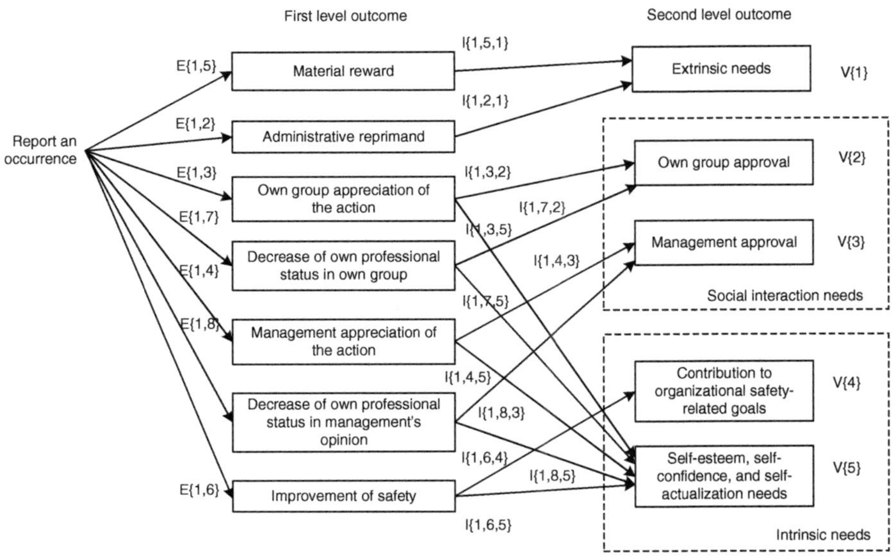

Fig. 6.2 Decision making model for reporting an occurrence

In the Vroom's model the force on an individual to perform an act is defined as:

$$\mathrm{F_i} = \sum_{\mathrm{j=1}}^{\mathrm{n}} \mathrm{E}\{\mathrm{i,j}\} \cdot \sum_{\mathrm{k=1}}^{\mathrm{m}} \mathrm{V}\{\mathrm{k}\} \times \mathrm{I}\{\mathrm{i,j,k}\}$$

Here E{i,j} is the strength of the expectancy that act A{i} will be followed by outcome j; V{k} is the valence of the second level outcome k; I{i,j,k} is perceived instrumentality of outcome j for the attainment of outcome k for act i. The action alternative with the highest force is chosen to be performed by the human.

The agent's decision making consists in the evaluation of the forces for two alternatives: to report and to not report. The agent chooses to perform the alternative with a greater force. In the following the basis for calculation of the variables of the decision making model for reporting is discussed (Fig. 6.2). The precise, elaborated details of the mathematical model can be found in (Sharpanskykh et al. 2008; Sharpanskykh and Stroeve 2011).

The factors E{1,5}, E{1,2}, I{1,5,1} and I{1,2,1} are defined based on the ANSP's formal reprimand/reward policies (see Table 6.1). In particular, E{1,2} = 1 for an observed occurrence, which completes a set of occurrences, for which a reprimand is defined; E{1,2} = 0 for all other observed occurrences. The values of E{1,3} and I{1,3,2} depend largely on the average commitment of the team of controllers to safety, and E{1,8} and I{1,4,3} depend on the management commitment to safety.

With each set of occurrences, in which a controller agent was involved during an evaluation period (e.g., a month), the measure of severity is associated, calculated as the sum of the severities of the occurrences from the set. The factors E{1,7}, E{1,8}, I{1,7,2}, I{1,4,3} depend mostly on the severity of the set of occurrences of the controller known to his/her team and known to the management. E{1,6} is based on the agent's beliefs about the dependencies between previous reporting of similar occurrences and improvement of safety that followed.

E{1,7}, I{1,3,5} and I{1,7,5} are based on the agent's IDV index, which indicates the degree of importance of group's opinions for the agent. I{1,4,5} and I{1,8,5} are based on the agent's PDI index. Furthermore, also the values of the basis valences (the degrees of importance of particular needs taken alone, see Fig. 6.2) of a controller agent depend on its cultural indexes (inspired by (Pinder 1998; Hofstede 2005), and empirical data):

$$v1_b = 1 \; v2_b = 1 - \text{IDV} \; v3_b = 0.7 \cdot \text{PDI} + 0.3 \cdot \text{UAI} \; v4_b = 0.3 + 0.7 \cdot \text{UAI}$$

The values of valences change over time depending on the degree of satisfaction of the agent's needs: the more a need is satisfied, the less its valence:

$$v(need) = \begin{cases} v_b \cdot \dfrac{\min\_accept(need)}{sat(need)}, sat(need) \geq \min\_accept(need) \\ v_b + v_b \cdot \dfrac{\min\_accept(need) - sat(need)}{\min\_accept(need)}, sat(need) < \min\_accept(need) \end{cases}$$

here sat(need) is the current satisfaction value of a need.

6.5 Safety Culture Modeling

Based on the cognitive and behavioral states described in the previous sections and in the complete model specification (Sharpanskykh et al. 2008; Sharpanskykh and Stroeve 2011) precise expressions for the safety culture indicators identified in the introduction have been determined.

SCI1: Average reporting quality of controllers

$$\text{SCI1} = \sum\nolimits_{a \in \text{CONTROLLER}, ot \in \text{OCC\_TYPE}} n_{ot,a} / (|\text{CONTROLLER}| \cdot |\text{OCC\_TYPE}| \cdot m_{ot,a}),$$

where $n_{ot,a}$ is the number of occurrences of type ot observed and reported by an agent controller a; $m_{ot,a}$ is the number of occurrences of type ot observed by an agent controller a; CONTROLLER is the set of all names of the controller agents; OCC_TYPE is the set of all names of the occurrence types.

SCI2: Average quality of the processed notification reports

$$SCI2 = \sum_{a \in CONTROLLER, t \in TIME, ot \in OCC\_TYPE, o \in OCCUR} e15_{o,ot,a,t} / |OCCUR|,$$

where $e15_{o,ot,a,t}$ is the quality of the notification report for an occurrence o of type ot reported by an agent a at time point t; OCCUR is the set of names of all occurrences.

SCI3: Average quality of the final safety occurrence assessment reports

$$SCI3 = \sum S / |S|,$$

where S = {v | ∃a:CONTROLLER, o:OCCUR, cl:OCC_TYPE, t:TIME has_state(a, belief(final_report(occurrence(o, cl, a), v), t))}, has_state(a, s) indicates that agent a has state s, final_report(occurrence(o, cl, a), v) indicates the quality v of the final report for the occurrence o of severity class cl of agent a.

SCI4: Average quality of monthly safety overview reports received by controllers

$$SCI4 = \sum S / |S|,$$

where S = {v | ∃a:CONTROLLER, m:MONTH, t:TIME has_state(a, belief (monthly_report(m, v), t))}

SCI5: Average commitment to safety of controllers

$$SCI5 = \sum_{a \in CONTROLLER, t \in TIME} e6_{a,t} / (|CONTROLLER| \cdot |TIME|),$$

Here $e6_{a,t}$ is the commitment to safety of an agent a at time point t; CONTROLLER is the set of all names of controller agents, and TIME is the set of all time points.

SCI6: Average commitment to safety of management as perceived by controllers

$$SCI6 = \sum_{a \in CONTROLLER, t \in TIME} e2_{a,t} / (|CONTROLLER| \cdot |TIME|),$$

Here $e2_{a,t}$ is the perception of the commitment to safety of management of agent a at time point t, CONTROLLER is the set of all names of the controller agents, TIME is the set of all time points.

6.6 Experimental Results

The developed generic model was instantiated for the following ANSP types in the Western and Eastern European cultural contexts (see also Table 6.5):

Table 6.5 The organizational aspects of the ANSP types used in simulation, classified along the scale (L)ow – (A)verage – (H)igh

ANSP type organizational aspect	*1*	*2*	*3*	*4*	*5*	*5a*	*6*	*6a*	*7*
Formal commitment to safety	H	H	H	H	H	H	A	L	L
Investment in personnel	A	A	H	H	H	H	L	L	L
Quality of technical systems	A	A	H	H	H	H	A	A	A
Formal support for confidentiality of reporting	A	A	H	H	H	H	L	L	L
Quality of management of safety activities	L	L	H	L	H	H	L	L	L
Personal consequences of occurrences	H	A	L	L	H	H	H	H	L
Personal rewards for reporting	–	H	–	–	–	–	–	–	–
Influence of a controller on organizational safety arrangements	L	L	H	H	A	H	L	L	L
Quality of identification of occurrences	A	L	L	L	H	H	H	H	L

1. Organization with a high formal safety commitment, but a lower actual commitment. Organization performs average control over activities of controllers and reprimands for occurrences.
2. Organization with a high formal safety commitment, but a lower actual commitment. The control over activities of controllers is low and reprimands are given only for serious occurrences. Rewards are provided for reporting series of less severe occurrences.
3. Formally committed organization, which puts substantial investments in safety. No reprimands are provided for occurrences, except for the class A.
4. Formally committed organization, which puts substantial investments in safety. However, the quality of management of safety activities is low. No reprimands are provided for occurrences, except for the class A.
5. Formally committed organization which puts substantial investments in safety. Organization performs close control over activities of controllers and reprimands for occurrences.

5a. Formally committed organization which puts substantial investments in safety. Organization performs close control over activities of controllers and reprimands for occurrences. Influence of controllers on safety arrangements is high.

6. Organization with a high formal safety commitment, but a lower actual commitment. Organization performs close control over activities of controllers and reprimands for occurrences.

6a. Organization has low commitment to safety and makes low investment in safety. Organization performs close control over activities of controllers and reprimands for occurrences.

Table 6.6 The simulation results (means) for the safety culture indicators of an Eastern European ANSP (E) and of a Western European ANSP (W)

Safety culture indicator (SCI)		*1*	*2*	*3*	*4*	*5*	*5a*	*6a*	*6*	*7*
1. Reporting quality	E	0.67	0.82	0.86	0.85	0.77	0.77	0.45	0.46	0.20
	W	0.77	0.87	0.9	0.9	0.77	0.78	0.43	0.49	0.37
2. Quality of notification reports	E	0.44	0.53	0.73	0.72	0.67	0.67	0.24	0.26	0.11
	W	0.5	0.56	0.77	0.76	0.67	0.68	0.23	0.27	0.20
3. Quality of safety occurrence assessment reports	E	0.19	0.18	0.5	0.33	0.51	0.5	0.07	0.07	0.07
	W	0.18	0.19	0.5	0.34	0.5	0.5	0.07	0.08	0.07
4. Quality of monthly safety overview reports	E	0.48	0.49	0.86	0.69	0.86	0.86	0.33	0.33	0.32
	W	0.48	0.48	0.86	0.69	0.86	0.86	0.33	0.34	0.33
5. Commitment to safety	E	0.60	0.65	0.74	0.72	0.7	0.73	0.37	0.48	0.36
	W	0.5	0.5	0.68	0.67	0.61	0.68	0.37	0.43	0.37
6. Perceived commitment to safety of management	E	0.54	0.55	0.79	0.73	0.74	0.77	0.25	0.38	0.25
	W	0.55	0.54	0.77	0.73	0.74	0.77	0.25	0.39	0.25

7. Organization has low commitment to safety and makes low investment in safety. No reprimands are provided for occurrences, except for the class A, and control over the activities of controllers is not strict.

Relations between the qualitative labels from Table 6.5 used for specifying the organizational types and corresponding numerical values of the model variables are provided in (Sharpanskykh et al. 2008). The instantiated models were populated with 48 controllers agents distributed over 6 air traffic control sectors, working in 4 homogeneous shifts, 12 h per day during 3 years (12 controllers per shift; 2 per sector). Based on each model, 1,000 simulations were performed. The purpose of the simulation experiments was to evaluate the quality of safety culture w.r.t. safety occurrence reporting depending on the organizational settings in different types of organizations. In each simulation the values of the safety culture indicators identified in the introduction were calculated in the Eastern and Western European cultural contexts (see Table 6.6).

It follows from the simulation results that the formal reward/reprimand system of an ANSP has a noticeable impact on reporting. In particular, the results for setting 5a versus setting 3 show that the introduction of reprimands and of a close control over activities of controllers in the ANSP's that are committed to safety, causes a notable decrease in the reporting quality in both cultures. On the contrary, it follows from a comparison of the results of setting 6a versus setting 7, that in organizations with little commitment to and investments in safety there is a significant increase in reporting quality as results of reprimands and a close control over controller agents. In such organizations these measures thus could be considered as instruments to make controller agents report (forcedly).

It can be observed in Table 6.6 that in general, the controllers in the Eastern European ANSP are more sensible to changes of the organizational settings than the controllers in the Western European ANSP. This is reflected in larger variation of the reporting quality among different organizational settings in the Eastern European case. It may be explained by a higher dependence of the Eastern European controller agents on the behavior and values of their peers within a shift and in relation to the management.

The quality of the processed notification reports produced depends considerably in both cultures on the quality of technical systems used by controllers and on investment in personnel. Furthermore, in the Western European ANSP the degree of influence of controllers on organizational safety arrangements has a notable impact on the quality of the processed notification reports (settings 1, 2, 5). Moreover, in both cultures the controllers tend to decrease the quality of notification reports (e.g., by holding back from informing some relevant details) in the conditions of high personal consequences of occurrences. The lowest quality of notification reports occurs in simulations of an Eastern European ANSP with a low commitment to and investment in safety and in personnel.

The quality of the feedback (received final safety occurrence assessment report) is almost the same for both cultures. Furthermore, according to the simulation, the quality of the feedback is not a determining factor for the controller's decision to report; this factor has an effect when it is combined with other factors (for example, as in settings 6 and 7).

The quality of the received monthly safety overview reports depends in both cultures mostly on the investment in personnel and on the quality of management of safety activities. Also, a positive correlation between the reporting quality and the quality of the received monthly safety overview reports can be observed in the simulation results.

As can be seen from the simulation results, the controller's commitment to safety in both cultures is influenced greatly by the perceived actual organizational commitment to safety. Furthermore, quality of management of safety activities has little impact on the controller's commitment in both cultures. The controller's commitment to safety in the Western European culture is influenced notably by the perceived controller's influence on organizational safety arrangements (for example, see setting 3). The commitment of controllers in the Eastern European ANSP is influenced by the ANSP's reward/reprimand system and by the quality of identification of occurrences (for example, see setting 2), whereas a similar dependence has not been found for the Western European ANSP. The organizational commitment to safety has a greater impact on the safety commitment of controllers in the Eastern European culture than in the Western European culture. The perceived commitment to safety of management is almost the same for the controllers in both cultural contexts.

6.7 Conclusions

James Reason once said that 'Few phrases occur more frequently in discussions about hazardous technologies than safety culture. Few things are so sought after and yet so little understood.' (Reason 1997). Although currently a considerable amount of work has been done to characterize safety culture via survey studies, the causal relations with organizational processes performed in a particular national cultural context are in general still vague. This chapter proposes an approach to systematically develop models that account for a large variety of organizational aspects, thus providing a different and structured view on safety culture from the perspective of the formal organization in relation with the variable behavior of agents in it. Such modeling provides the opportunity of the structured development of policies for improvement of safety culture. The development of the model has been done on the basis of data from existing ANSPs. The obtained simulation results provide remarkable insights in potential relations between the quality of occurrence reporting and organizational factors at an ANSP. As was demonstrated by simulation, some of these relations depend on the national culture, which is also supported by existing literature (Choudhry et al. 2007; Ek et al. 2007).

The model contains a large number of parameters and relations between variables. To determine the influence of changes in parameter values on the model's outputs an extensive sensitivity analysis study was performed, which results are reported in (Stroeve and Sharpanskykh 2009; Stroeve et al. 2011).

Safety culture professionals from EUROCONTROL recognized a high potential of the proposed approach. In particular, the approach may further enhance safety culture questionnaires and be used to prepare safety culture survey workshops.

Another promising direction for future research is to integrate the developed approach into existing agent-based approaches such as (Tumer and Agogino 2007) and (Blom et al. 2006), which address systems at the operational level and do not consider the organizational and cultural contexts. In such a way, explicit reciprocal relations between organizational structures, processes and safety culture on the one side and operational safety indicators (e.g., risks) on the other side can be defined. The identification of such relations enables understanding and profound analysis of the system behavior at different aggregation levels (e.g., individual, team, organization).

References

Blom, HAP, Stroeve, SH, De Jong, HH (2006). Safety risk assessment by Monte Carlo simulation of complex safety critical operations. In Redmill F, Anderson T (eds.), Developments in risk-based approaches to safety, Springer-Verlag, London, 47–67

Carver, CS, Scheier, MF (2000) Perspectives on Personality. Needham Heights, MA: Simon & Schuster.

Choudhry RM, Fang D, Mohamed S (2007) The nature of safety culture: A survey of the state-of-the-art. Safety Science 45:993–1012

Ek, A, Akselsson, R, Arvidsson, M, Johansson, CR (2007) Safety culture in Swedish air traffic control. Safety Science, 45(7): 791–811

Goldberg, LR (1993) The structure of phenotypic personality traits. American Psychologist, 48: 26–34

Hersey, P, Blanchard, KH, Johnson, DE (2001) Management of Organizational Behavior: Leading Human Resources. Prentice-Hall, Inc., New Jersey.

Hofstede, G (2005) Cultures and Organizations. McGraw-Hil.

Hopkins, A (2006) Studying organizational cultures and their effects on safety. Safety Science, 44: 875–889

Pinder, CC (1998) Work motivation in organizational behavior. NJ: Prentice-Hall

Popova, V, Sharpanskykh, A (2008) Process-oriented organisation modelling and analysis. Enterprise Information Systems Journal, 2(2):157–176

Reason J (1997) Managing the risks of organizational accidents. Ashgate, Aldershot, England

Sharpanskykh, A (2008) On Computer-Aided Methods for Modeling and Analysis of Organizations, PhD thesis, VU University Amsterdam

Sharpanskykh, A, Stroeve, SH (2008) Safety modelling and analysis of organizational processes in air traffic, D3: Methods and requirements. National Aerospace Laboratory NLR, report NLR-CR-2008-299.

Sharpanskykh A, Stroeve SH, Van Lambalgen R (2008) Safety modelling and analysis of organizational processes in air traffic – D4: Model and initial simulation results. National Aerospace Laboratory NLR, report CR-2008-407

Sharpanskykh, A, Treur, J (2010) A Temporal Trace Language for Formal Modelling and Analysis of Agent Systems. In: Specification and Verification of Multi-Agent Systems (book chapter), Springer Verlag, 317–352

Sharpanskykh, A, Stroeve, SH (2011). An agent-based approach for structured modeling, analysis and improvement of safety culture. Computational & Mathematical Organization Theory 17(1):77–117

Stroeve SH, Sharpanskykh A (2009) Safety modelling and analysis of organizational processes in air traffic – D7: Validation. National Aerospace Laboratory NLR, report CR-2009-305

Stroeve, S, Sharpanskykh, A, Kirwan, B (2011) Agent-based organizational modelling for analysis of safety culture at an air navigation service provider. Reliability Engineering & System Safety, 96:515–533

Trompenaars, F, Hampden-Turner, C (1997) Riding the Waves of Culture: Understanding Cultural Diversity in Business. London: Nicholas Brealey

Tumer, K, Agogino, A. Distributed Agent-Based Air Traffic Flow Management, In:6th International Joint Conference on Autonomous Agents and Multiagent Systems (AAMAS 2007), Honolulu, Hawaii (2007)

Uttal, B. The corporate culture vultures. Organizational Culture Journal, 17, 66–72 (1983)

Chapter 7
Monolingual Biases in Simulations of Cultural Transmission

Seán Roberts

7.1 Introduction

Cultural groups are very rarely isolated. They interact for trade, politics and war. Communication is key to these interactions, and so a common language is important. The emergence of common languages has been studied using computational models. However, one aspect of cultural interaction has been left largely ignored – the ability to learn many languages at once, or bilingualism. This chapter considers the importance of incorporating bilingualism into studies of cultural evolution.

Bilingualism is by no means a rare phenomenon. Statistics on the exact prevalence of bilingualism are difficult to obtain. In the USA 18 % of the population are estimated to speak two or more languages (U.S. Census Bureau 2003). The estimate is 34 % for Canada (Statistics Canada 2007), 66 % in the EU (Euopean Comission 2006) and 80 % in China (Baker and Jones 1998). Bilinguals are a majority in about a third of countries (Baker and Jones 1998). These are likely to be conservative estimates, and with over 6,000 languages squeezed into in around 200 nations, it's likely that contact with multiple languages is an everyday feature of most people's lives.

Recently, industrialisation and globalisation have meant that, in the first world, the perception of the prevalence of bilingualism is artificially low – especially for native speakers of global languages such as English (Thomas and Wareing 1999; Kostoulas-Makrakis 2001; Luchtenberg 2002). It's no surprise, then, that when cultural processes come to be modeled, one of the first simplifying assumptions would be that people speak one language. However, the abstraction to monolingualism ignores several linguistic phenomena such as the prevalence of bilingualism in societies and the ease with which children learn more than one language (Pearson et al. 1993).

S. Roberts (✉)
Language and Cognition Group, Max Planck Institute for Psycholinguistics,
Wundtlaan 1, 6525 XD Nijmegen, The Netherlands
e-mail: sean.roberts@mpi.nl

V. Dignum and F. Dignum (eds.), *Perspectives on Culture and Agent-based Simulations*, Studies in the Philosophy of Sociality 3, DOI 10.1007/978-3-319-01952-9_7,

This chapter will consider the validity of monolingual assumptions in models of cultural evolution. Firstly, the way in which bilingualism might affect the evolutionary dynamics of language is explored. Next, a case-study of the 'Minimal Naming Game' will reveal an implicit monolingual bias, namely mutual exclusivity (the assumption that each object only has one name and each name only refers to one object, see Markman and Wachtel 1988 and Merriman and Bowman 1989). Since bilinguals do not exhibit mutual exclusivity (Byers-Heinlein and Werker 2009; Healey and Skarabela 2009; Houston-Price et al. 2010), the model is generalised to weaken this constraint. The model demonstrates that communicative success can be achieved even without mutual exclusivity, in opposition to previous research (Smith 2002, 2009). The model suggests that cultural phenomena adapt to the function they are required to fulfill (e.g. Christiansen and Chater 2008 and Beckner et al. 2009). When seeking to model the integration of cultures a common measurement is required. However, even small differences in the way different communities interact can lead to fundamental cultural differences between them, meaning that a common metric might be very abstract.

7.1.1 Bilingualism and Cultural Evolution

The dynamics of language evolution have been extensively studied through computational modelling. The canonical language learner in these models is an agent that tries to settle on a single grammar that explains the variation in its input. This implicit monolingualism is seen as a necessary abstraction in order to get at the more fundamental dynamics of language evolution. There is a sense in the field of language evolution that bilingualism is a sociolinguistic phenomenon that is the product of the interactions of several monolingual communities who have already evolved language. Implicitly, bilingualism is seen as a secondary linguistic ability – a sort of by-product.

For instance, many models represent languages as discrete entities which compete with one another (Niyogi and Berwick 1995; Abrams and Strogatz 2003). Even when language is modelled as distributions over words, two standard simplifying assumptions are made by many approaches to language evolution and change (e.g. Griffiths and Kalish 2007; Kirby 2001 and Smith et al. 2003). Firstly, it is assumed that there are discrete generations with one agent per generation. This limits the amount of complexity that can be added by the cultural system. Secondly, it is assumed that all learners use the same learning algorithm, or that learning algorithms do not change over a learner's lifetime.

The first assumption has already been criticised (Niyogi and Berwick 2009; Burkett et al. 2010) and recent research has shown that the complexity of cultural dynamics can effect the eventual distribution of languages in a population (Smith 2009). A model has also been proposed which allows agents to speak and acquire multiple languages from multiple speakers (Burkett et al. 2010).

However, the second assumption may also be called into question. I will illustrate this with research on the mutual exclusivity bias, and continue in the next section to show that this bias exists in certain models of language evolution and change. It has been demonstrated that monolingual children and adults exhibit a mutual exclusivity bias (Markman and Wachtel 1988; Merriman and Bowman 1989): a tendency to assume that each object only has one name and each name only refers to one object. However, recent research has shown that bilinguals do not exhibit mutual exclusivity (Byers-Heinlein and Werker 2009; Healey and Skarabela 2009; Houston-Price et al. 2010). It is hypothesised that the bias is overridden because of a higher variance in the input of children in bilingual contexts. Applying mutual exclusivity when presented with two languages is not suitable, since there will be at least two words for each object.

If the amount of linguistic variance (at any level of description) influences the learning strategy for that variance, then this will affect the selective pressure on languages. This will, in turn, affect the kinds of languages that emerge, thus feeding back into the amount of linguistic variance. These aspects would then co-evolve.

Given this, there are two possible fundamental states of the language learner. Either they begin with a mutual exclusivity bias which is overridden in certain situations or they begin with no assumptions and develop mutual exclusivity if the conditions are right. In the next section, it will be shown that some models make implicit assumptions about the development of mutual exclusivity and see it more as a fundamental part of language acquisition and language evolution rather than an acquired heuristic that is applied in suitable contexts. It will be argued that the most abstract learner is one without the mutual exclusivity bias, and so models should not assume mutual exclusivity as part of the learner's bias.

7.2 Categorisation Games

This section presents a case-study of a model of cultural evolution – the Categorisation Game – and demonstrates implicit monolingual biases that obscure some interesting dynamics. The Categorisation Game looks at how agents in a population converge on a shared system for referring to continuous stimuli (Nowak and Krakauer 1999; Steels 1996; Steels and Belpaeme 2005). This paradigm is often couched in terms of deciding on words for objects referred to by their colour. The colour spectrum is continuous, so agents must decide where to place category boundaries as well as the label for that category. The 'minimal naming game' (Loreto et al. 2010) (also used in Gong et al. 2008; Puglisi et al. 2008 and Baronchelli et al. 2010) is a simplification of the categorisation game which "possibly represents the simplest example of the complex processes leading progressively to the establishment of complex human-like languages" (Loreto et al. 2010). I'll show that even this 'minimal' algorithm has implicit monolingual assumptions. First, however, a note is made about the measurements that researchers have used to study the categorisation game.

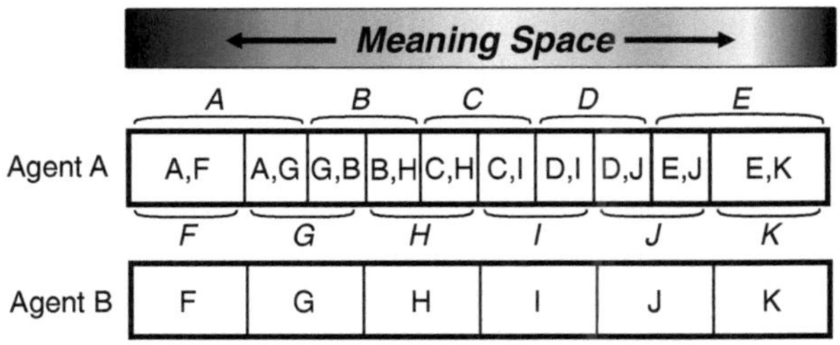

Fig. 7.1 An example of how two agents might split the meaning space into categories and label those categories. The meaning space spans the interval 0.0–1.0. *Agent A* and *B* both have the same conceptual space, but *agent A* has multiple labels in each category while *agent B* only has one label in each category. The representation for *agent A* above pulls apart sections of the space that are contiguously labelled with the same label into two systems

7.2.1 Measurements of Coherence

Other models looking at this problem have considered measurements apart from communicative success. For instance, the 'level of lexical coherence' in the system, according to Baronchelli et al. (2006) is the average proportion of shared lexical items in a population. The category overlap function (Loreto et al. 2010; Puglisi et al. 2008) measures the level of alignment between the category boundaries of the agents. However, an appropriate measurement when considering the possibility of 'bilingualism' is less clear. For instance, consider the example of two agents with categories and labels as described in Fig. 7.1. Adapting the lexical coherence measurement from Baronchelli et al. (2006) gives a coherence of 75 %. This measurement fails to capture the fact that agent B would always be understood by agent A and that agent A could always make itself understood to agent B given the right choice of lexical item. In other words, although the agents have differences in the words that they know, they are still able to communicate unambiguously about the whole spectrum.

Measuring category overlap is also problematic. Agents with category boundaries at exactly the same locations will have a category overlap of 1.0. However, the overlap of the example above is 0.09, despite the relatively good communicative success possible between the pair. This is because the measurement collapses the category boundaries of an agent into a single system before comparing it to another agent. By doing this, the division between the two 'languages' of agent A in Fig. 7.1 is ignored.

These measures reflect the level of coherence in the population, but only effectively for a population whose goal it is to converge on a single, 'monolingual' system. Researchers have used these measurements to gauge the progress of their model, demonstrating a monolingual bias in their approach. Further research is required to find a good way of measuring coherence in a heterogenous population (see Komarova and Jameson 2008; De Vylder 2006 and De Beule 2006). This paper will proceed assuming that communicative success should be the most important measure of coherence between agents.

7.2.2 The Minimal Naming Game

The algorithm for the categorisation game is reproduced below. However, two of the steps are re-analysed as heuristics rather than essential elements. These heuristics impose a mutual exclusivity bias in the agents. The steps are as follows (following Puglisi et al. 2008): There is a population of N agents, each able to partition the perceptual space into categories. Each category has a list of associated words. Each agent has a minimum perceptual difference threshold d_{min}, below which stimuli appear the same. At each time step:

1. Two individuals are chosen at random to be the speaker and the listener.
2. They both have access to a scene containing M stimuli. The stimuli must be perceptually distinguishable by the agents (perceptual distance $\geq d_{min}$).
3. The speaker selects a topic and discriminates it in the following way:

   - Each stimulus is assigned to a perceptual category
   - If one or more other stimuli are assigned to the same category as the topic, the agent splits its perceptual categories so that each stimulus belongs to only one perceptual category. Within a category with two or more stimuli, a boundary is placed halfway between the first two stimuli.
   - The new partitions inherit the associated words of the old partition.
   - **Heuristic A**: Each new partition is given a new, unique name. It's assumed that no two agents will create the same name.

4. The speaker transmits a word that it associates with the topic to the listener. If it has no words associated with the category, it creates a new one. If it has more than one word associated, it transmits the one that was last used in a successful communication.
5. The hearer receives the word and finds all categories which have the associated word and which identify one of the stimuli in the scene. Then:

   - If there are no such categories, the agent does nothing.
   - If there is one such category, the agent points to the associated stimulus.
   - If there is more than one such category, the agent points randomly at an associated stimulus.

6. The hearer discriminates the scene, as above.
7. The speaker reveals the topic to the listener.
8. If the hearer did not point to the topic, the communication is a failure. The hearer adds the transmitted word to the category discriminating the topic.
9. If the hearer pointed to the topic, the communication is a success.
   Heuristic B: Both agents delete all other words but the transmitted one from the inventory of the category discriminating the topic.

Heuristic A, above, invents new words for each sub-category when a category is split. This is an implementation of the assumption that each name only refers to one object, hence when there are two objects with the same name, the agent should discriminate between them linguistically. This interacts with Heuristic B which removes all competing names associated with a category from the listener's lexicon when communication is successful. The effect is that the listener conforms to the speaker's labeling, but also 'forgets' any previously associated words. This is an implementation of the assumption that each object only has one name.

These two heuristics, then, implement a mutual exclusivity bias: Each name only refers to one object and each object is only labeled by one name. Stable bilingualism is impossible in this model because only one name is retained after successful communication. The role of the two heuristics in the evolution of a shared communicative system is clear: heuristic A creates new labels for categories, introducing variation into the system needed to distinguish between categories. Heuristic B causes the agents to converge on shared labels for categories by selecting for labels common to an interacting pair.

However, these heuristics are still arbitrary. As we have seen, not all human learners assume mutual exclusivity. In the next section, it will be demonstrated that a population of agents can converge on a shared communication system without these heuristics.

7.3 Convergence Without Mutual Exclusivity

The algorithm was modified to remove the mutual exclusivity bias in order to test the effects on communicative success. However, the changes to the dynamics will not be explored in detail. The purpose of the changes, here, is not to explore the best way of modelling the cultural evolution of language, but to demonstrate that the biases of the researcher can influence the dynamics of the model and thus the conclusions drawn from it.

Heuristic B can be modified while retaining communicative success (Baronchelli 2011). If the hearer, but not the speaker applies heuristic B, a coherent vocabulary still emerges in a similar time with similar memory resources required. If only the speaker applies heuristic B a coherent vocabulary does emerge, but on a longer timescale and in a qualitatively different way (approached as a thermodynamic system, consensus is reached due to large, system-size fluctuations of the magnetisation (Baronchelli 2011)). However, this research was concerned with the effect of feedback on the convergence dynamics. This study looks at the assumptions built in to the individual learning algorithm.

The heuristics were modified by generalising the algorithm. Firstly, agents in a population either all applied heuristic A or all did not apply heuristic A. Heuristic B was made optional in the same way. If heuristic B did not apply, a maximum number of words s_{MAX} were retained after a successful communication. A first-in, first-out stack memory was also implemented so that the oldest stored form would be removed first. A word was pushed further back in the stack (safer from deletion) when a listener heard it being used by a speaker. This is a generalisation of the mechanism that weakens links between signals and meanings which do not co-occur.

The purpose of generalising the model was to allow bilingualism. However, the advantages of knowing more than one word for an object are not yet fully available. A bilingual, failing to communicate with one word, might try another. Therefore, the algorithm was modified to allow an arbitrary maximum number of attempts a_{MAX} at communicating before communication failed. If speakers had more than one label for a perceptual category, they transmitted them in a random order until this maximum was reached. Listeners searched their lexicon at each attempt until either they found a match in their own lexicon and made a guess at the referent or the maximum number of attempts was reached and they signaled failure, as before. Each guess was independent of any other, so successful communication was not always guaranteed, even when $a_{MAX} = M$.

It has been shown that an algorithm which leads to successful communication in a population of agents must strengthen connections between signals and meanings that appear together (or are absent together) and weaken connections between signals and meanings that do not co-occur (Smith 2002). The changes to the algorithm above do not violate these conditions, but simply weaken their strength.

7.4 Results

Four versions of the algorithm were run: with both heuristics, as in the original, with only heuristic A, only heuristic B and with neither heuristic. Results shown here are for a population of 4 over 10,000 rounds with a context size of 2.

7.4.1 Communicative Success

Table 7.1 and Fig. 7.2 show the communicative success for the algorithm run with different heuristics with $a_{MAX} = 2$ and $s_{MAX} = 2$. All heuristics manage in achieving good communicative success at some point (shown by the maximum communicative success achieved). That is, a mutual exclusivity bias is not necessary for communicative success in this model. It should be noted that the probability of choosing the correct referent by chance is $\frac{1}{c} = 0.5$ (where c is the context size) because the algorithm tends to limit the number of words linked to a perceptual category to one. However, algorithms without heuristic B (i.e. 'A only' and 'None') have a higher probability because they are more likely to be able to take advantage of extra communicative attempts. Therefore, for algorithms without heuristic B, the probability of selecting the correct referent by chance is

$$\sum_{i=1}^{a_{MAX}} \left(\frac{1}{c}\right)^i \tag{7.1}$$

Table 7.1 Communicative success for different heuristics for 10 runs each with 4 agents and $a_{MAX} = 2$ and $s_{MAX} = 2$ for 10,000 rounds. The maximum communicative success was calculated as follows: for each run, the average communicative success for data grouped into 100-round bins. The average of a bin was taken as the value of that bin. The values shown in the table are the average of the maximum bin values for 10 runs. The final communicative success is the average success for last 100 rounds over 10 runs. The statistics show a two tailed t-test indicating performaance above chance

Heuristics	Max	t	p	Final	t	p
A and B	0.84	23.89[a]	< 0.0001	0.62	3.06[a]	0.01
A only	0.95	44.26[b]	< 0.0001	0.86	7.17[b]	< 0.0001
B only	0.77	9.04[a]	< 0.0001	0.52	1.44[a]	0.18 (N.S.)
None	0.94	43.50[b]	< 0.0001	0.84	5.96[b]	< 0.001

[a]Calculated with chance level at 0.5
[b]Calculated with chance level at 0.75

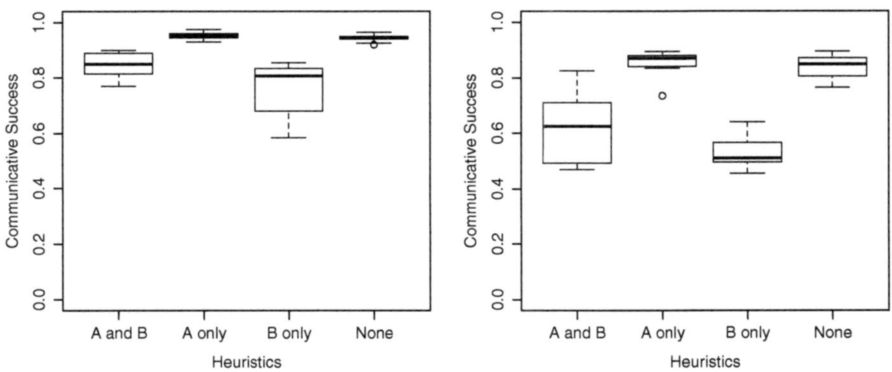

Fig. 7.2 Maximum communicative success (*left*) and final communicative success after 10,000 rounds (*right*) for 10 runs of populations with various heuristics. Statistics in Table 7.1

For the current settings, this is 0.75. Even taking this into account, all algorithms are able to reach stable periods with high levels of communicative success. The result is robust against changes to s_{MAX}: The relative communicative success between the different heuristic combinations remains the same for s_{MAX} up to 1,000, while the absolute communicative success drops about 5 % for s_{MAX} of 4 and remains around that level for s_{MAX} up to 1,000.

However, eventually all agents converge on a single word for the whole meaning space. This is typical behaviour for this model (Baronchelli 2006). This reduces the communicative success, since agents cannot distinguish linguistically between referents. Table 7.1 shows the average final communicative success after 10,000 rounds. These are less than the maximum. In the case of using heuristic B only, the communicative success is no better than chance. The other algorithms still yield a communicative success above chance, but the algorithms without heuristic B (A only and no heuristics) do better than algorithms with heuristic B (average with B = 0.57, without B = 0.85, $t = 10.9$, $p < 0.0001$). The same collapsing process

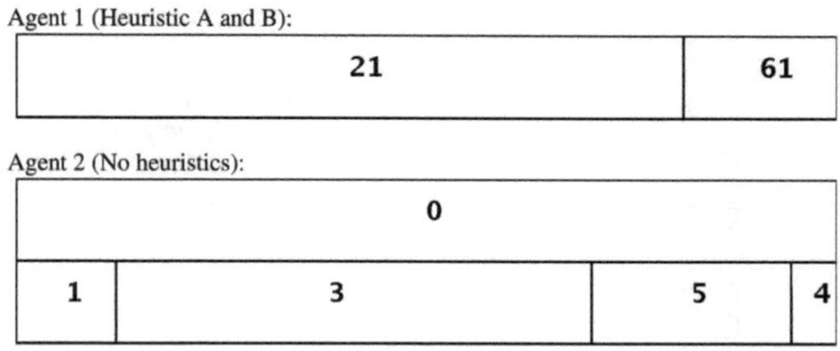

Fig. 7.3 The linguistic labels of an agent after 3,500 rounds for separate runs. Agent 1 (*above*) used heuristics A and B and agent 2 (*below*) used neither heuristic. The perceptual space runs from left to right. Contiguous linguistic categories are indicated by boxes with the linguistic label (a number in this implementation) drawn in the centre. Agent 2 has more than one label for a given perceptual stimulus

occurs as in the algorithm without heuristic B, but since there is more variation within perceptual categories due to extra labels being stored, a single label takes longer to dominate. In fact, a single linguistic item tends to spread over the whole meaning space as with the original algorithm, but a sort of secondary 'language' keeps distinctions between perceptual categories for longer.

Figure 7.3 illustrates this with a diagram of agents' memories from mid-way through separate runs. Agent 1 was run in a population using both heuristics and agent 2 was run in a population using neither heuristic. Agent 1's linguistic categories are already heavily collapsed while Agent 2 has a greater variation which allows it to communicate more effectively. The memories of both agents at this point are nearly perfectly similar to the other agents that they interact with.

Another measure of communicative efficiency is the entropy efficiency of an agent. Effectively, this is the average probability that an agent has a different linguistic label for any two stimuli. An agent has a set of linguistic labels which uniquely identify regions of the meaning space. L is the list of lengths of these regions. The entropy efficiency is given as

$$-\sum_{i=1}^{|L|} \frac{L_i \, log(L_i)}{log(1/d_{min})} \tag{7.2}$$

Since d_{min} is set so that there can be a maximum of 10 perceptually distinct regions, the highest entropy efficiency is given by an agent who can uniquely label 10 regions of equal length (entropy efficiency of 1.0). The lowest possible entropy efficiency is given by an agent with no labels or one label spanning the whole meaning space (entropy efficiency of zero). Figure 7.4 shows that the algorithm with both heuristics achieves a lower entropy efficiency than the algorithm without heuristic B and degrades faster than the algorithm without heuristic A.

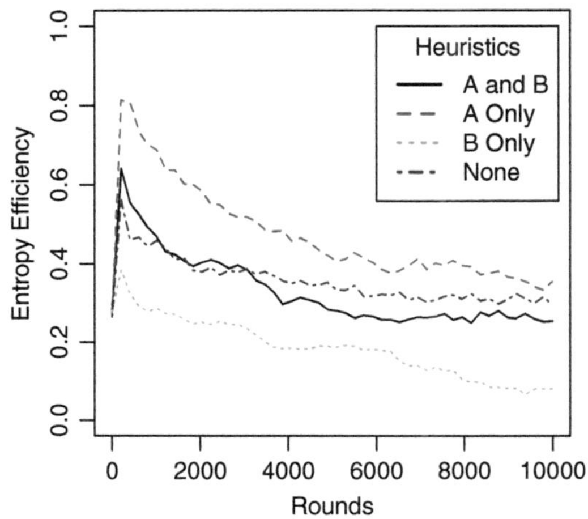

Fig. 7.4 Entropy efficiency for populations of agents with different heuristics. Number of agents = 4, $a_{MAX} = 2$ and $s_{MAX} = 2$

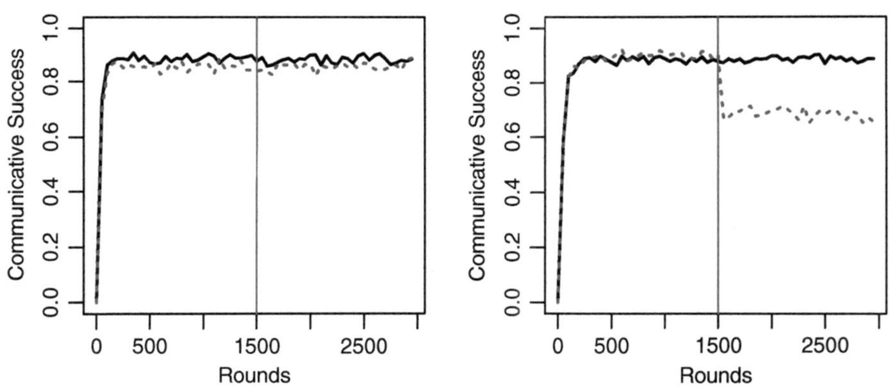

Fig. 7.5 Communicative success for a population of two agents (*left*) and a population of four agents (*right*). *Solid lines* indicate success for a consistent algorithm where no heuristics are applied. *Dashed lines* represent success for an algorithm that incorporates the heuristics after 1,500 rounds

7.4.2 The Development of Mutual Exclusivity

The model has shown that mutual exclusivity is not necessary for communicative success. However, the mutual exclusivity bias is exhibited by monolinguals. The model can be manipulated to explore the rationale behind this and the most likely starting assumptions of a language learner.

Simulations were run where the mutual exclusivity heuristics were 'switched on' after some rounds. Figure 7.5 shows the difference between an algorithm that has no heuristics and one that changes to incorporate them after 1,500 rounds. For

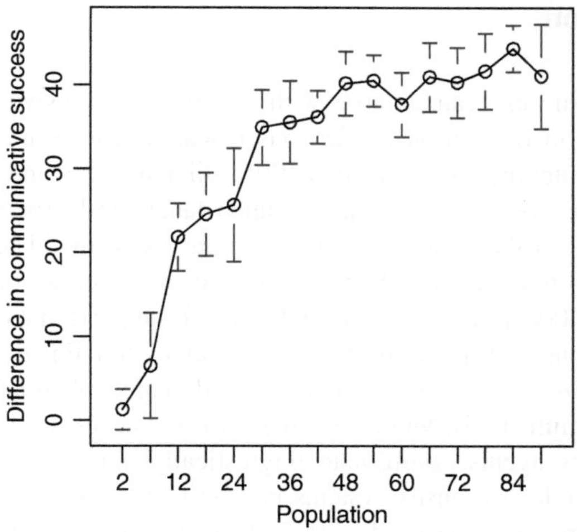

Fig. 7.6 The difference in percentage communicative success when switching from using no heuristics to using both heuristics for different population sizes (the difference between *solid* and *dashed lines* after 1,500 rounds in Fig. 7.5). Values shown are mean and 99 % confidence intervals for five runs. Positive values indicate an advantage for using no heuristics. In a population of 2, there is no difference between using the heuristics or not, as shown in Fig. 7.5

a population of two agents (low cultural complexity), switching on the heuristics makes no difference to the communicative success. Therefore, in this situation, applying mutual exclusivity makes rational sense in order to save memory: The application of heuristic B will reduce the number of words stored for each category. However, in a population of four agents, switching on the heuristics decreases the communicative success. In this situation, the most rational approach is to keep the heuristics switched 'off'. This is because the complexity of the cultural system is greater with four agents, leading to more variation between agents. The system evolves to store many words for an object to cope with this variation. The drop in this difference reflects the empirical findings that bilinguals do not exhibit mutual exclusivity. Figure 7.6 shows that this difference increases with larger populations. However, when s_{MAX} becomes many times greater than the number of agents, the disadvantage of switching decreases. That is, agents retain words that have already been discarded by others.

The most rational strategy for any agent is not to assume mutual exclusivity to begin with, and only to activate it under relevant conditions. This reflects the findings that 14-month-old children do not exhibit it while 17-month-olds do (Halberda 2003). From this model we might conclude that mutual exclusivity is an acquired heuristic which is applicable in situations where there is likely to be low variation (monolingualism). More research is required into this kind of model. The point here is that the assumptions of the original model obscure the distinction between mutual exclusivity as an innate, universal bias and an acquired, culture-specific one.

7.5 Discussion

Communicative success can emerge without mutual exclusivity. The results of this model stand in opposition to previous research (e.g. Smith 2002; Vogt and Haasdijk 2010; Hutchins and Hazlehurst 1995; Oliphant 1999 and De Vylder 2006). For instance, it has been claimed that "human language learners appear to bring a one-to-one bias to the acquisition of vocabulary systems. The functionality of human vocabulary may therefore be a consequence of the biases of human language learners" (Smith 2004, p. 127). The current research suggests that mutual exclusivity is not an innate bias. Furthermore, the bias becomes functional as a consequence of the variance in the vocabulary and social dynamic. A related model shows similar results (Smith 2005): Mutual exclusivity is not necessary for communicative success, but helps agents co-ordinate linguistically when they have conceptual differences. Multiple consensus systems can be maintained in a population with complex social structures (de Vylder 2007). However, the current model shows that mutual exclusivity does not always aid the co-ordination process.

However, rather than directly opposing the claims of some previous models, the constraints in the current model can be seen as a relaxation of the constraints embodied by the mutual exclusivity bias. Both models contribute the necessary ingredients for an evolutionary system: Heredity, variation and differential fitness (e.g. Lewontin 1970). Although generational turnover is not modelled, there is heredity in the sense that each agent inherits its own memory from the previous round. Heuristic A introduces the variation by adding new words. Heuristic B introduces differential fitness by selecting words which are successful in communication. Without heuristic A, variation is still introduced by agents creating new words at early stages of the game when they have no words at all (step 4 of the algorithm). The generalisation of Heuristic B to keep an arbitrary number of words after successful communication allows selection to operate over *groups* of words rather than single ones.

Heuristics A and B, then, are an efficient way of introducing the ingredients for evolution into the system. However, cultural processes can also introduce these ingredients – the individual learning processes need not be the source. Other processes could also introduce variation such as errors in production or perception or differences in contact with other agents.

7.6 Conclusion

The naming game was reanalysed in the light of evidence from bilingual language acquisition research. The measurements used to analyse the model were also re-assessed and shown to favour monolingual systems. Steps in the categorisation game were re-analysed as implementing a mutual exclusivity constraint. To explore the effects of these steps, the learning algorithm was generalised so that the steps could be omitted. Communicative success at the lexical level was achieved without mutual exclusivity constraints. In fact, in some cases, the constraint impedes the process.

This goes against some previous research which argued that mutual exclusivity is necessary for communication to emerge. What seems to be important is the presence of the ingredients for evolution – inheritance, variation and selection. The mutual exclusivity bias is seen as an efficient way of integrating these ingredients. However, the model also showed that rational agents should not assume mutual exclusivity to begin with. This reflects research which shows that children only start using mutual exclusivity in certain situations. Mutual exclusivity is not appropriate in a bilingual environment, so bilinguals do not exhibit it. Given this, the monolingual assumptions of the naming game are unrealistic for two reasons. First, a learner's learning algorithm may change over time, as demonstrated by the differences found between monolinguals and bilinguals. Secondly, they are not valid abstractions because the heuristics which implement mutual exclusivity are optional extras, so the simplest, default assumptions of learners should be bilingual. That is, monolingualism is a specialised form of bilingualism.

When modelling cultural processes, abstraction is necessary. However, the cultural phenomena that appear simplest (e.g. monolingualism) may not be caused by the simplest learning mechanisms. Much of the complexity in cultural phenomena stem from complex interactions *between* individuals. That is, the cultural transmission process itself can shape and influence the cultural practices it transmits.

7.6.1 Integrating Cultures in the Light of Cultural Adaptation

The communication system in the model above adapts to fit the needs and constraints of its users. Indeed, the hallmark of a cultural phenomenon is that it has adapted to the cognitive niche of its community's members (Christiansen and Chater 2008; Beckner et al. 2009). If different communities have different dynamics, such as population size or differences in social structures, then the cultural phenomena that emerge in them may be radically different. In the model above, the communication system between two agents became optimised for efficiency while the communication system in a more complex social structure became optimised for flexibility. Biases in communities towards these different optimisations could be amplified by cultural transmission (Kirby et al. 2007). Over many generations, and for a more complex cultural phenomenon (e.g. a language system, judicial system or musical form), the commonalities between two communities may erode to very abstract principles. When seeking to integrate them, then, a common measure for separate cultures may be difficult to find. Even something as simple as assuming each object only has one associated word may reflect the deep structure of the culture in which it is embedded.

Acknowledgements Thank you to Simon Kirby, Kenny Smith, Antonella Sorace and Liz Irvine for comments. Supported by the Economic and Social Research Council [ES/G010277/1].

References

Abrams, D.M. and Strogatz, S. H.: Modelling the dynamics of language death. Nature, 424, pp. 900. (2003)

Baker, C. & Jones, S. P.: Encyclopedia of bilingualism and bilingual education. Clevedon, UK: Multilingual Matters. (1998)

Baronchelli, A.: Role of feedback and broadcasting in the naming game. Physical Review E. 83(4), 046103. (2011)

Baronchelli, A., Dall'Asta, L., Barrat, A. & Loreto, V.: Bootstrapping communication in language games in A. Cangelosi, A. D. M. Smith & K. Smith, eds.: The Evolution of Language, Proceedings of the 6th International Conference (EVOLANG6), World Scientific Publishing Company. (2006)

Baronchelli, A., Felici, M. Loreto, V., Caglioti, E. &Luc Steels:Sharp transition towards shared vocabularies in multi-agent systems. Journal of Statistical Mechanics: Theory and Experiment. P06014. (2006)

Baronchelli, A., Gong, T., Puglisi, A., and Loreto, V.: Modeling the emergence of universality in color naming patterns. Proceedings of the National Academy of Sciences, 107(6):2403–2407. (2010)

Burkett, D., Griffiths, T.: Iterated learning of multiple languaged from multiple teachers. In Smith, A., Schouwstra, M., de Boer, B., and Smith, K., editors, The Evolution of Language: Proceedings of EvoLang 2010, 58–65. (2010)

Byers-Heinlein, K. & Werker, J. F.: Monolingual, bilingual, trilingual: infants' language experience influences the development of a word-learning heuristic. Developmental Science, 12(9), 815–823. (2009)

Christiansen, M. H. & Chater, N.: Language as shaped by the brain. Behavioral and Brain Sciences, 31, 5, 489–508; discussion 509–58. (2008)

De Beule, J. and De Vylder, B. and Belpaeme, T. A cross-situational learning algorithm for damping homonymy in the guessing game. In L. M. Rocha, L. Yaeger, M. Bedau, D. Floreano, R. Goldstone, and A. Vespignani (eds.) Artificial Life X, 466–472. MIT Press. (2006)

De Vylder, B. & Tuyls, K.: How to reach linguistic consensus: A proof of convergence for the naming game. Journal of Theoretical Biology, 242 (4) 818–831. (2006)

de Vylder, B. The Evolution of Conventions in Multi-Agent Systems. PhD thesis, Artificial Intelligence Laboratory, Vrije Universiteit Brussels. (2007)

Euopean Comission. 2006. Special Eurobarometer: Europeans and their Languages: European Commission. http://ec.europa.eu/public_opinion/archives/ebs/ebs_243_en.pdf.cited3Nov2010.

Gong, T., Puglisi, A., Loreto, V., & Wang, W. S-Y.: Conventionalization of linguistic knowledge under simple communicative constraints. Biological Theory: Integrating Development, Evolution, and Cognition, 3 (2) 154–163. (2008)

Griffiths, T. L. & Kalish, M. L.: Language Evolution by Iterated Learning With Bayesian Agents. Cognitive Science: A Multidisciplinary Journal. 31 (3) 441–480. (2007)

Halberda, J.: The development of a word-learning strategy. Cognition, 87 (1) 23–34. (2003)

Healey, E. & Skarabela, B.: Are children willing to accept two labels for one object? A comparative study of mutual exclusivity in bilingual and monolingual children. In Proceedings of the Child Language Seminar. University of Reading. (2009)

Houston-Price, C., Caloghiris, Z., & Raviglione, E.: Language Experience Shapes the Development of the Mutual Exclusivity Bias. Infancy 15 (2), 125–150. (2010)

Hutchins, E., & Hazlehurst, B.: How to invent a lexicon: The development of shared symbols in interaction. In N. Gilbert & R. Conte (eds.) Artificial societies: The computer simulation of social life, pp. 157–189, London: UCL Press. (1995)

Lewontin, R. C.: The Units of Selection. Annual Reviews of Ecology and Systematics 1: 1–18. (1970)

Loreto, V., Baronchelli, A., and Puglisi, A.: Mathematical Modeling of Language Games, Evolution of Communication and Language in Embodied Agents, part 3, 263–281. Springer Berlin Heidelberg. (2010)

Luchtenberg, S.: Bilingualism and bilingual education and their relationship to citizenship from a comparative German-Australian perspective, Intercultural Education, Vol. 13(1), 49–61. (2002)
Kirby, S.: Spontaneous evolution of linguistic structure-an iterated learning model of the emergence of regularity and irregularity. IEEE Transactions on Evolutionary Computation, 5 (2) 102–110. (2001)
Kirby, S., Dowman, M. & Griffiths, T.:Innateness and culture in the evolution of language. Proceedings of the National Academy of Sciences, 104:12, 5241–5245. (2007)
Komarova, N. L. & Jameson, K.A. Population heterogeneity and color stimulus heterogeneity in agent-based color categorization.Journal of Theoretical Biology. 253(4): 680–700. (2008)
Kostoulas-Makrakis, N.: Language and Society. Basic Concepts (in Greek). Athens: Metexmio. (2001)
Markman, E. M. and Wachtel, G. A.: Children's use of mutual exclusivity to constrain the meanings of words. *Cognitive Psychology.* 20, 121–57. (1988)
Merriman, W. E. and Bowman, L. L.: The mutual exclusivity bias in children's word learning. *Monographs of the Society for Research in Child Development.* 54 (3–4). (1989)
Niyogi, P. & Berwick, R. C.: The Logical Problem of Language Change. CBCL Paper 115, MIT AI Laboratory and Center for Biological and Computational Learning, Department of Brain and Cognitive Sciences. (1995)
Niyogi, P. and Berwick, R.C.: The proper treatment of language acquisition and change in a population setting. Proceedings of the National Academy of Sciences. 106 (25) 10124–9. (2009)
Nowak, M. A., & Krakauer, D. C.: The evolution of language. *Proceedings of the National Academy of Sciences of the United State of America.* 96 (14) 8028–8033. (1999)
Oliphant, M.: The learning barrier: Moving from innate to learned systems of communication. Adaptive Behavior, 7, 371–384. (1999)
Pearson, B.Z., Fernandez, S.C., Oller, D.K.: Lexical development in bilingual infants and toddlers: Comparison to monolingual norms. Language Learning, 43, 93–120. (1993)
Puglisi, A., Baronchelli, A. & Loreto, V.: Cultural route to the emergence of linguistic categories *Proceedings of the National Academy of Science.* 105 7936–7940. (2008)
Smith, K: The cultural evolution of communication in a population of neural networks. Connection Science, 14, 65–84. (2002)
Smith, K.: The evolution of vocabulary. Journal of Theoretical Biology. 228 (1) 127–142. (2004)
Smith, A. D. M.: Mutual exclusivity: communicative success despite conceptual divergence. In M. Tallerman (ed.) Language Origins: perspectives on evolution. 372–388. Oxford University Press. (2005)
Smith, K.: Iterated learning in populations of Bayesian agents. Proceedings of the 31st Annual Conference of the Cognitive Science Society. (2009)
Smith, K., Kirby, S. & Brighton, H.: Iterated learning: a framework for the emergence of language. Artificial Life 9 (4) 371–86. (2003)
Statistics Canada: The Evolving Linguistic Portrait, 2006 Census Statistics Canada Catalogue no. 97–555-XIE. Ottawa. Version updated May 2010. http://www12.statcan.ca/census-recensement/2006/as-sa/97-555/pdf/97-555-XIE2006001-eng.pdf (3/11/2010). (2007)
Steels, L.: Perceptually grounded meaning creation in M. Tokoro (Ed.) *Proceedings of the International Conference on Multi-agent Systems*. MIT Press. Cambridge, MA. (1996)
Steels, L. and Belpaeme, T.: Coordinating perceptually grounded categories through language: a case study for colour. Behavioral and Brain Sciences, 28(4):469–489. (2005)
The "Five Graces Group": Beckner, C., Blythe, R., Bybee, J., Christiansen, M. H., Croft, W., Ellis, N. C., Holland, J., Ke, J., Larsen-Freeman, D. & Schoenemann, T. Language Is a Complex Adaptive System: Position Paper. Language Learning, 59, 1–26. (2009)
Thomas, L., & Wareing, S.: Language, Society and Power. London: Routledge. (1999)
U.S. Census Bureau: Language Use and English-Speaking Ability: 2000 (C2KBR-29). Washington: Government Printing Office. http://www.census.gov/prod/2003pubs/c2kbr-29.pdf (3/11/2010). (2003)
Vogt, P. & Haasdijk, E.:Modeling social learning of language and skills. Artificial Life 16 (4) 289–309. (2010)

Part III
Culture Simulation

Chapter 8
Towards Agent-Based Models of Cultural Dynamics: A Case of Stereotypes

Jens Pfau, Yoshihisa Kashima, and Liz Sonenberg

8.1 Introduction

Culture has been studied from two different perspectives: as a relatively stable system of meaning shared by a group of people, and as processes of meaning making that people engage in (Kashima 2000). While the former macro-level perspective underlies most cross-cultural comparative research, the latter micro-level perspective facilitates the understanding of the situated expression and acquisition of cultural information. However, ultimately it is micro-level interactions that people engage in that give rise to the macro-level distribution of cultural information. Likewise, the macro-level distribution of information affects how, when, and whether this information is communicated at all in a given situation. Therefore, studying the interaction between micro- and macro-level is necessary for understanding *cultural dynamics*, namely, the formation, maintenance, and transformation of culture over time.

A prominent contemporary meta-theoretical approach that attempts to examine cultural dynamics is neo-diffusionism (Boyd and Richerson 1985; Cavalli-Sforza and Feldman 1981; Dawkins 1976), the view that takes culture as the socially transmitted information prevalent in a group of people. In this view, the critical issue is the process of social transmission of information between individuals. Information transmitted frequently at the micro-level becomes shared within a group and part of this group's culture at the macro-level.

A closer look at the actual process of cultural transmission reveals that it involves a highly interactive and joint activity. In particular, the *grounding model of cultural transmission* by Kashima et al. suggests that cultural transmission is often an incidental consequence of everyday joint activities (Kashima et al. 2008). The

---

J. Pfau (✉) • Y. Kashima • L. Sonenberg
The University of Melbourne, Parkville, VIC 3010, Australia
e-mail: jpfau@unimelb.edu.au; ykashima@unimelb.edu.au; l.sonenberg@unimelb.edu.au

V. Dignum and F. Dignum (eds.), *Perspectives on Culture and Agent-based Simulations*, Studies in the Philosophy of Sociality 3, DOI 10.1007/978-3-319-01952-9_8,

model postulates that the transmission of cultural information is a function of the purpose of the joint activity, and of the common ground that does exist between the participants and the common ground that they perceive to exist. To capture these highly dynamic, complex, and interactive processes, we propose that agent-based models provide an appropriate tool for the study of cultural dynamics. In *agent-based models*, individuals represented by autonomous computer programs interact according to programmatically defined rules, mimicking social interactions (Gilbert 2008). The experimenter can intervene both at the micro-level by changing the interaction rules of individuals as well as on the macro-level by changing how a condition, such as the distribution of a particular piece of information, affects micro-level interactions. Thereby, a detailed understanding of a large-scale social system can be gained even when the behavior of this complex dynamical system becomes analytically intractable.

There are several lines of work on agent-based models of cultural dynamics. Similarly to prevailing neo-diffusionist theories, these models typically assume that cultural transmission is a matter of simple imitation, relying on epidemic dynamics and abstracting from the micro-process of cultural transmission (Goldstone and Janssen 2005; Maxwell and Carley 2009). For instance, most prominent is Axelrod's model of the dissemination of culture and the large body of research that has built on his work (Axelrod 1997). Much research has also been conducted by the statistical physics community that studies social systems in terms of particles (Castellano et al. 2009). As successful as these models are in reproducing empirically observed macro-level patterns, they do not represent the complexity of human interaction and communication as joint activities. Therefore, these models cannot support the search for a psychological theory of cultural dynamics that bridges the gap between micro- and macro-levels.

Accordingly, we propose to develop agent-based models of cultural dynamics which take seriously the micro-dynamics of interpersonal interaction and communication. Building on the grounding model of cultural transmission, we claim that two elements need to be present in any detailed agent-based model of cultural dynamics: (1) the representation of the joint activity that agents are engaged in and their incidental task-oriented dialogue, as well as (2) the dependency of this communication on the agents' common ground. We propose a semi-formal model that integrates these two aspects by building on a computational model of joint activity and a discourse protocol that enables the exchange of task-relevant information while abstracting away from the level of utterances. Our model is a more explicit description of the grounding model of cultural transmission, thus serving as a framework for agent-based models that represent joint activities as the engine of cultural dynamics. Thereby we contribute to understanding the interrelation between micro- and macro-level of cultural dynamics.

We rely on stereotypes as an example of cultural information. Existing computational models of stereotype formation and change in the social psychological literature have mostly investigated the intra-personal processes of the storage and recall of memories, without considering the agent's cognition to be embedded in social interactions let alone a larger social context, (e.g. Kashima et al. 2000; Queller and Smith 2002; Smith and DeCoster 1998 and Van Rooy et al. 2003). Not until recently a perspective of cognition as a socially situated process has been applied to the modeling of stereotypes (Smith and Collins 2009; Van Overwalle and Heylighen 2006; Van Rooy 2009).

We summarize the grounding model of cultural transmission by Kashima et al. in the next section. In Sect. 8.3, we sketch a translation of the model into concepts from intelligent agent research. We conclude the chapter by outlining the impact of this work on the development of agent-based models of cultural dynamics, and by making suggestions for future research.

8.2 The Grounding Model of Cultural Transmission

In this section, we sketch Kashima et al.'s grounding model of cultural transmission which postulates that cultural information is mainly transmitted during everyday joint activities (Kashima et al. 2008). We discuss concepts and processes that are relevant for this chapter and treated more formally in the next section. In particular, we describe how the process of grounding is an implication of joint activities and how it is affected by the context of the activity and the existing common ground.

A joint activity can be as simple as having small-talk with a co-worker but it can also consist of multiple, hierarchically organized sub-activities. Two or more actors are involved in any joint activity, contributing by their individual actions which are regulated by their intentions. However, these intentions obviously need to be coordinated. Accordingly, philosophers have theorized about joint activities in terms of *joint intentions* or *we-intentions*, for example, the intention that we write a paper together, distributing writing work but aligning content, style and language, and proofreading each other's contributions. Some argue that joint intentions can be reduced to individual intentions and mutual beliefs (Bratman 1992) while others disagree (Searle 1990; Tuomela 2006). We do not commit yet to any particular theory, but note that participants need to properly intend to perform their parts of the joint activity in coordination with their partners.

In order to coordinate, participants need to communicate and align their beliefs about information that is relevant to the successful execution of the activity. Building on Clark's theory of grounding in language use (Clark 1996), we apply the term *grounding* to describe this alignment process. Grounding thus is a subordinate process to the participants' joint activity.

Clark postulates that during communication interlocutors rely heavily on their *common ground*. He (Clark 1996, p. 95) defines common ground as follows:

Definition 8.1 (Common Ground according to Clark). A proposition, ϕ, is common ground for members of a group G if and only if: (i) the members of G have information that ϕ and that (i).

When interlocutors begin to engage in their joint activity, they start with a certain initial common ground, which is due to their previous shared experience (*personal common ground*) or their group membership (*communal common ground*). Subsequent grounding during the interaction adds new information to their common ground. Grounding consists of at least two different phases: (1) the presentation of some information, say, a proposition ϕ by an interlocutor, and (2) the acceptance by its partner, which signals that the speaker's intent with regard to the presentation of ϕ has been understood. However, the listener's acceptance is in fact itself a presentation of the proposition that the listener has understood the speaker's presentation. It then needs to be accepted by the speaker. Thus, in principle, the presentation-acceptance pair can continue indefinitely; however, when the interlocutors regard a certain proposition as mutually understood to the extent sufficient for the current purpose (as defined by the joint activity), they stop the presentation-acceptance exchange, and treat it as common ground. As part of this exchange, both interlocutors can request clarifications from each other when they cannot understand their partner sufficiently for the current purpose. The proposition eventually accepted by all interlocutors is added to their common ground. Thus, common ground is constructed collaboratively, and the proposition eventually added to common ground is not necessarily the proposition ϕ that the speaker intended originally.

Clark (1996) developed his model of grounding in order to explain language use at the utterance-level. That is, his theory is about how interlocutors establish a sufficient basis that they have understood an utterance, but not about whether they agree with the communicated content or not. We assume that if a proposition implied by an utterance is presented and accepted by interlocutors, it is encoded by the interlocutors as true, unless there is an explicit denial of the truthfulness of this proposition (Gilbert et al. 1990, 1993). Likewise, we assume that unless the proposition implied by an utterance is explicitly disagreed or questioned, the interlocutors would regard it as mutually agreed with.

We call the common ground created during a particular joint activity *context-specific common ground*. Context-specific common ground is indexed by the time and location of the activity as well as the identities of the participants. However, context-specific common ground can be generalized temporally, spatially, or socially: Interlocutors usually assume that grounded information actually constitutes context-specific common ground for the next interaction, and should be mutually accessible again if the interaction continues at another location or time. Likewise, interlocutors can infer that their context-specific common ground is actually shared by a wider community. They can also infer that a proposition grounded for a particular person applies to a group of people. These processes of generalization link personal to communal common ground.

We assume that not only *target information* is grounded during an interaction, but also *presuppositional* and *relational information*. Target information is information that is explicitly grounded; presuppositional information is information presupposed by the target information; and relational information concerns the social relationship between the interlocutors or with other individuals or groups implied by the target information. This information is individually inferred to be part of common ground.

Hence, interlocutors might come to different views of what their common ground is and we need to distinguish between *actual common ground* and *perceived common ground*.

We assume that although the joint activity largely dictates which information needs to be communicated, and therefore determines *epistemic goals*, the joint activity typically implies certain *relational goals* as well, that is, goals of regulating social relationships among the interlocutors. Epistemic goals are managed by generic strategies such as Grice's (1975) communication maxims or Sperber's and Wilson (1995) *principle of relevance*. Relational goals can be managed, for example, by Levinson's (1983) politeness rules. However, these goals might be incompatible at times, thus posing a dilemma: Sharing information that is accurate (epistemic goal) but inconsistent with common ground might require more effort during the grounding process, which might have an adverse effect on the relationship between the interlocutors (Clark and Kashima 2007). Modifying such information so that it is more consistent with common ground might lead to a smoother grounding process and might hence be socially-connective (satisfying a relational goal), but may also amount to the dropping of some content. Interlocutors need to manage what they communicate in order to achieve these possibly competing goals.

Stereotypes are an instance of cultural information and hence their transmission follows similar rules to the ones defined above. For the purpose of this chapter, it shall be sufficient to understand a stereotype as a (possibly commonly held) generalized belief about a social group. As *stereotype-relevant* information we consider information that has a relationship with this belief, either because it supports or contradicts it. From the perspective of the grounding model of cultural transmission, stereotypes and stereotype-relevant information can be part of common ground and thus play a role in the grounding process. For example, information that is consistent with stereotypes is preferred to be communicated over information inconsistent with stereotypes if the stereotype is perceived to be shared within the interlocutors' community (Lyons and Kashima 2003).

In this section, we have outlined the grounding model of cultural transmission to an extent sufficient for us to proceed with a more formal treatment in the following. In particular, we have built on Clark's model of grounding in language use to describe the transmission of cultural information as an implication of joint activity. We have discussed how this process depends on the context of the interaction and the existing common ground. A more comprehensive treatment of the grounding model of cultural transmission can be found in Kashima et al. (2008).

8.3 Towards a Formalization

In this section, we work towards a formal representation of the grounding model of cultural transmission. We rephrase the processes described before by building loosely on concepts from intelligent agent research. We do not commit to any particular formalization of these notions yet but rather rely on their intuitive meaning. In particular, we first discuss the basic ingredients for a model of computational intelligent agents on which our account is based. Thereafter, we identify how a joint activity and the agents' participation can be described. Subsequently, we describe our formalization of common ground, the process of grounding and its interrelation with epistemic and relational goals. Then we sketch a formalization of the interrelation between the joint activity and the grounding process. We also discuss briefly how the temporal, spatial, and social generalization of cultural information can be described in this framework.

We are not addressing the construction of a full-fledged dialogue system but a model that describes the alignment of mental models during joint activities. That is, we do not address the utterance-level of communication but a higher level at which information about beliefs and concepts is exchanged. In doing so, we mainly rely on the observation discussed previously that if a proposition implied by an utterance is presented and accepted by interlocutors, it is encoded as true, unless its truthfulness is explicitly questioned.

8.3.1 The Agent Model

We assume that an agent engaging in a joint activity has a set of *beliefs* that is updated by internal (reasoning) and external events (perception). By $Bel_A(\phi)$ we denote that agent A believes proposition ϕ. A belief ϕ is called a *mutual belief* ($MB_G(\phi)$) of a group of agents G if all members of G believe ϕ and all believe that ϕ is a mutual belief. Note the correspondence with Clark's recursive definition of common ground in Sect. 8.2. More intuitively, a proposition ϕ is mutual belief among a group of agents G if all members believe ϕ, all members believe that all believe ϕ, all members believe that all believe that all believe ϕ, and so on ad infinitum. We assume that the agent generates *goals*—world-states that it would like to bring about (e.g. "I would like that we have this paper finished."). We denote by $Goal_A(\phi)$ that agent A has adopted ϕ as a goal. From all possible goals, the agent selects a subset to be pursued actively. However, only goals that are deemed achievable and compatible with each other and with the currently active set of goals can be selected. The agent is said to be committed to its set of active goals and will engage in planning activities to achieve these goals. Commitments are assumed to be binding: The agent will not drop any selected goals arbitrarily.

From here on we will call adopted goals *intentions* and we distinguish two different types (Grosz and Kraus 1996): An agent has an *intention-that* if it is committed to bring about a certain state of affairs (e.g. "I intend that I will have finished this paper by tomorrow."). An agent has an *intention-to* if it is committed to performing a certain action (e.g. "I intend to write the discussion section today."). In the following, we denote agent A's intention-that by $Int.Th_A(\phi)$ where ϕ is a proposition. An intention-to is denoted by $Int.To_A(a)$ where a is an action. Even though an intention-that does not imply any behavior directly, it can cause planning activities and lead to the adoption of further intentions-to. We assume that the *plan* or *recipe* selected or created for achieving the execution of an intention-to can in fact evoke the adoption of further subgoals. Note that an agent can have an intention-that whose target involves another agent but an intention-to can have as a target only the subject of this intention. Further discussions about formal models of intelligent agent systems can for example be found in Wooldridge (2009).

8.3.2 The Joint Activity

We discussed previously that a joint activity can be described as, or is driven by, a joint intention. For now, we will be agnostic about an exact definition assuming only that this intention describes a joint goal. However, we will rely on some commonly accepted properties of joint activity (Bratman 1992):

Mutual responsiveness Collaborating agents are trying to be responsive to each other's intentions and actions while knowing that the other party is doing so as well.

Commitment to the joint activity The agents are committed to the joint activity, which causes their mutual responsiveness. The reasons (individual intentions) why the agents are committed to the activity, however, do not need to be the same.

Commitment to mutual support The agents are committed to helping each other in order to complete the joint activity successfully.

We assume that the joint activity between agents A and B can be described as some sort of a joint intention-that which fulfills the above listed requirements and is held by both agents: $Int.Th_{\{A,B\}}(\phi)$ where ϕ is the proposition to bring about. In fact, ϕ could refer to the execution of actions of the participating agents. Because of the commitment entailed by the joint intention, agents will engage in planning activities to achieve their joint goal. In contrast to the achievement of individual intentions, however, communication might be necessary to coordinate planning and execution. Therefore, enabling individual intentions cannot be adopted without consideration of the partner's activities.

It is the activity that dictates which information might need to be exchanged, and it is the properties of mutual support presented above that cause agents to identify such information needs. When an information need arises, an agent will either present information to its partners that it deems necessary for them to fulfill their part of the task, or it will actively seek information that it requires itself. Similarly, the receiving agent will answer any request or acknowledge its understanding and agreement of presented information. We assume that any such communication attempt will induce a subordinate grounding process, whose implicit goal is for the agents to align their personal task-relevant information.

To illustrate how our model is able to describe the exchange of cultural and in particular stereotype-relevant information, we rely on three sample dialogues.[1] The background is that Alice, an employee of the city's football club, was made aware that Gary, one of the club's players, was caught drink-driving the night before. Even though all dialogues are based on this same prior event, their contexts (common ground, epistemic and relational goals) differ, thus leading to vastly different outcomes.

The day after the incident, Alice and her work colleague Maria work on the problem of managing the reputation of some players including Gary's. Their overarching joint goal is to work out how to improve Gary's reputation. Gary's identity is in their common ground but the incident from last night is not. Thus, the information that Gary was caught drink-driving is highly relevant to the joint goal and therefore contributed by Alice according to her epistemic goals. Relational goals—the improvement of Alice's and Maria's relationship—play only a minor role, if at all.

```
(1)         Alice: Unfortunately Gary got caught
                   drink-driving just yesterday.
(1.1)       Maria: What happened? I didn't hear
                   about that.
(1.2)       Alice: He emptied a bottle after he heard of
                   his grandma's death but then decided
                   to visit his grandfather.
(1.2.1)     Maria: I didn't know he was that close
                   to them.
(1.2.2)     Alice: His parents were out of town quite
                   often because of their jobs and his
                   grandparents looked after him then.
(1.2.2.1) Maria: So he actually is a decent fellow?
(1.2.2.2) Alice: Yes it seems.
(2)         Maria: Oh. This incident really is bad luck
                   for him.
```

Later Alice communicates with her husband Bob about her day. The joint goal is to have a casual conversation. However, relational goals are not important because their relationship is already strong. Alice assumes that Gary's identity is part of their common ground as well as the stereotype that a majority of football players does not drink wine. Therefore, according to epistemic goals, the information that Gary got drunk would be redundant with their stereotypes and hence irrelevant. However, the information that he drank wine is relevant because novel to Bob.

[1]Note that while our formalism is not an attempt at describing the production of low-level utterances, we artificially construct dialogues whose utterances do fit the abstract level of mental alignment we address here.

```
(1)       Alice: Did you know that Gary is a wine
                 drinker?
(1.1)     Bob:   Who is Gary?
(1.2)     Alice: One of our players, the one that we
                 met the other day at Jimmy's.
                 You remember?
(1.2.1) Bob:     Mh, no. But ... a football player
                 drinking wine?
(1.2.2) Alice:   Some of them seem to like wine.
(2)       Bob:   Mh.
```

Another day Alice has a conversation with her casual acquaintance Stacy. Again, the joint goal is to have a casual conversation. Relational goals are strong because the women are only casual acquaintances. Likewise, their personal common ground is small. Transmitting information that is assumed to be consistent with stereotypes in communal common ground contributes to relational goals. Therefore, Alice tells Stacy that Gary was caught drunk-driving although it is not particularly novel and hence does not contribute to epistemic goals.

```
(1)     Alice: Gary ..., a player from our club, got
               caught by the police the other day.
(1.1) Stacy: What happened?
(1.2) Alice: The usual story. He got drunk, drove
               his Porsche at 150, and abused the
               police when he got caught. He ended up
               in jail for the night.
(2)     Stacy: These football players are all the same.
```

8.3.3 Common Ground and Stereotypes

The consequence of grounding is the change of the agents' common ground—the information that they believe to be shared and mutually believed to be shared. Considering the correspondence of Clark's definition of common ground in Sect. 8.2 with the recursive definition of mutual belief in Sect. 8.3.1, we describe *actual common ground* by mutual belief. That is, a proposition ϕ is actually in the common ground of agents A and B iff it is mutually believed:

Definition 8.2 (Actual Common Ground).

$$CG_{\{A,B\}}(\phi) \Leftrightarrow MB_{\{A,B\}}(\phi) \quad (8.1)$$

However, common ground is not an objective entity external to the agents' minds as discussed earlier. Instead, each agent has its own view of their common ground and these views can potentially differ. Therefore, we consider a proposition ϕ to be part of agent A's *perceived common ground* with agent B iff agent A believes that ϕ is mutually believed between them:

Definition 8.3 (Perceived Common Ground).

$$CG^A_{\{A,B\}}(\phi) \Leftrightarrow Bel_A(MB_{\{A,B\}}(\phi)) \quad (8.2)$$

In the case of our sample dialogues, Alice falsely assumes common ground with Bob for the identity of Gary but correctly identifies that Gary's identity is not common ground with Stacy. However, Gary's identity is both actual as well as perceived common ground between Alice and Maria.

Because it is not relevant to the purpose of this chapter, we assume that the two definitions above encompass both personal and communal common ground. In fact, Alice's employment by the football club is personal common ground between Alice and Bob. Bob, on the other hand, assumes that his stereotype about football players typically not drinking wine is part of communal common ground, that is, he expects this view to be commonly accepted. We denote a stereotype, a generalized belief ϕ about individuals x in a social group G by logical implication: $\forall x[G(x) \Rightarrow \phi(x)]$. A stereotype can be part of a community's communal common ground such that we would consider it as commonly held.

8.3.4 *The Grounding Process*

As indicated before, a grounding process is initially triggered when the agents' commitment to their joint task causes them to engage in dialogue for the purpose of exchanging task-relevant information. In this subsection, we are going to explore how this process could be modeled formally within the framework of joint activity described in Sect. 8.3.2.

The purpose of grounding is to reach a common understanding of the information communicated by the speaking agent A. We denote this information as a proposition ϕ, which for example could be a proposal for the receiving agent to adopt a certain belief or to carry out a certain action. We assume that the presentation of information by one agent and its subsequent acceptance by the other agent makes this information mutually believed, thus adding it to their common ground. We base this assumption on the observation discussed in Sect. 8.2 that people encode exchanged information that remains unchallenged as true. We denote these actions by *present*(ϕ) and *accept*(ϕ).[2] The *accept*-action could actually correspond to an implicit acceptance in real dialogue, for example by the dialogue moving on without any further challenge.

[2]We disregard the actual communication language and its formal semantics here to facilitate comprehensibility.

However, depending on whether the presented information is compatible with the listener's prior beliefs, an *accept*(ϕ) might not be immediately possible. In this case, the listening agent will object and request clarification from the speaker. The *accept*-action is not performed until this request is served. In fact, any clarification request or any clarification itself needs to be agreed on by the other agent and can therefore lead to yet another level of clarification. We denote these actions as *req-clarify*$_\omega(\psi, \theta_0)$ and *clarify*$_\omega(\theta_1)$ where ψ is some explanation offered by the listener why it cannot accept ω, θ_0 is an optional transformation on ω acting as a counter-proposal, and θ_1 is a clarifying transformation on θ_0 provided by the speaker. For example, in utterance 1.2.1 of the second sample dialogue, Alice adds the information to her original proposition that Gary is a football player in her club. Both θ_0 and θ_1 are mappings from the set of all possible propositions Φ to itself $\theta_i : \Phi \rightarrow \Phi$ and they could be substituting part or the whole of ω or add additional content. If a θ is the identity (I), it does not apply any transformation to its argument. As indicated already, $req\text{-}clarify.(\psi, \theta)$ can be a request to clarify the ϕ of a *present*(ϕ), or the θ of a *clarify*($\cdot, \theta$), or even the ψ or θ of a *req-clarify*.(ψ, θ). A *req-clarify* thus initiates a new sub-dialogue, which is completed successfully when either the listening agent issues the *accept*-action after any open *req-clarify* has been answered, or when its superordinate dialogue is completed, or when the *clarify* is issued and dialogue returns to the superordinate level.

Either of the agents can issue a *cancel* at any level of this recursive process of clarifications. This will terminate the entire grounding process. However, when finally ϕ has been transformed such that it can be accepted by the listener, the transformation of ϕ negotiated during the process will be grounded. Additionally, we assume that any transmitted and unchallenged information is assumed to be mutually believed and hence grounded, possibly transformed by any θ's. The information incidentally grounded at any point in time during the dialogue is exactly the one that is not referenced by any open sub-dialogue anymore. Thus *req-clarify* and *clarify* play a similar role in grounding as *present* and *accept*.

Take the second dialogue presented in Sect. 8.3.2 as an example. An illustration of what is happening in this dialogue is presented in Fig. 8.1. The upper part specifies some of the mental attitudes that Alice and Bob are holding at the beginning of this interaction: They have a joint intention to have a conversation (Pre 1) and both have the epistemic goal to transmit information relevant to this joint activity (Pre 2). We omit relational goals here because they are likely to play a minor role, considering that Alice and her husband are certainly close already. Alice and Bob have in their common ground the stereotype that football players do not drink wine (Pre 3). Alice, moreover, falsely assumes that the identity of Gary and his occupation are also in their common ground (Pre 4). Apart from that, Alice has some information about Gary's drinking (Pre 5).

The second part of the figure maps the sample dialogue onto the model of the grounding process described in this section. Subscripts on the actions determine which proposition the sub-dialogue at this level is addressing. For example, utterance 1.1 and 1.2 amount to a clarification of ϕ, while 1.2.1 and 1.2.2 are concerned with a clarification of θ_1. Utterance 1 corresponds to the presentation

(Pre 1)	$Int.Th_{\{Alice,Bob\}}(have\text{-}conversation)$	
(Pre 2)	$Int.Th_{Alice}(be\text{-}relevant) \wedge Int.Th_{Bob}(be\text{-}relevant)$	
(Pre 3)	$CG_{\{Alice,Bob\}}\{\forall x[football\text{-}player(x) \Rightarrow \neg drinks\text{-}wine(x)]\}$	
(Pre 4)	$CG^{Alice}_{\{Alice,Bob\}}\{who(Gary) \wedge football\text{-}player(Gary)\}$	
(Pre 5)	$Bel_{Alice}(drinks\text{-}wine(Gary) \wedge got\text{-}drunk(Gary) \wedge got\text{-}caught(Gary) \wedge \ldots)$	
(1)	A: $present(\phi)$	$\phi = drinks\text{-}wine(Gary)$
(1.1)	B: $req\text{-}clarify_{\phi}(\psi_0, \theta_0)$	$\psi_0 = \neg Bel_{Bob}(who(Gary))$ $\theta_0 = I$
(1.2)	A: $clarify_{\phi}(\theta_1)$	$\theta_1(\theta_0(\phi)) = \phi \wedge football\text{-}player(Gary)$
(1.2.1)	B: $req\text{-}clarify_{\theta_1}(\psi_2, \theta_2)$	$\psi_2 = Bel_{Bob}(\forall x[football\text{-}player(x) \Rightarrow \neg drinks\text{-}wine(x)])$ $\theta_2 = I$
(1.2.2)	A: $clarify_{\theta_1}(\theta_3)$	$\theta_3(\theta_2(\theta_1(\theta_0(\phi)))) = \theta_2(\theta_1(\theta_0(\phi))) \wedge \exists x[football\text{-}player(x) \wedge drinks\text{-}wine(x)]$
(2)	B: $accept(\phi)$	
(Post 1)	$+CG_{\{Alice,Bob\}}\{drinks\text{-}wine(Gary) \wedge who(Gary) \wedge football\text{-}player\text{-}1(Gary) \wedge \forall x[football\text{-}player\text{-}1(x) \Rightarrow football\text{-}player(x) \wedge drinks\text{-}wine(x)] \wedge \forall x[football\text{-}player\text{-}2(x) \Rightarrow football\text{-}player(x) \wedge \neg drinks\text{-}wine(x)]\}$	
(Post 2)	$-CG_{\{Alice,Bob\}}\{\forall x[football\text{-}player(x) \Rightarrow \neg drinks\text{-}wine(x)]\}$	

Fig. 8.1 An illustration of the grounding process in the second sample dialogue. The *upper part* of the figure lists the relevant part of Alice's and Bob's mental states before the dialogue. The *middle part* provides a trace of the grounding process and the *last part* explicates the changes to Alice's and Bob's mental states as a result of the grounding. See text for more details

of the information that Gary drinks wine (ϕ). Alice opts for transmitting this part of the story because it is novel to Bob, given that it is incompatible with their common stereotype. However, Bob is unable to identify Gary and hence unable to accept Alice's proposition. He requests a clarification in 1.1. and signals that he does not know who Gary is (ψ_0). He does not make use of his opportunity to transform Alice's original proposal of ϕ (θ_0 is the identity). In 1.2, Alice seeks to answer Bob's request by adding to her original proposition the information that Gary is a football player (θ_1).[3] However, a football player drinking wine is not compatible with Bob's stereotypes. Therefore, he issues another clarification request in 1.2.1 providing as the reason for his misunderstanding his belief that football players do not drink wine (ψ_2). Again, he does not offer any counter proposal to Alice's proposition (θ_2 is the identity). Alice understands Bob's request and answers with a clarification in 1.2.2.

[3] We omit here any information that is not relevant for our discussion, for example, that Gary is playing for Alice's club.

She makes sense of the fact that Gary drinks wine despite the stereotype of football players as non-wine drinkers by realizing that some football players do drink wine (θ_3). This amounts to a sub-categorization of the original stereotype. In fact, another solution could have been to mark Gary explicitly as an exception of the stereotype. After this transformation to the original proposition, Bob is finally able to match the information with his beliefs and accept (utterance 2).

The third part of the figure identifies the change to common ground caused by this dialogue according to the grounding model described here. A plus in front of a proposition indicates that this proposition was added to common ground, while a minus denotes that it was removed from common ground. Alice and Bob ground the information that Gary drinks wine, who he is, that he plays football, and that there are two subgroups of football players: those who drink wine and those who don't ($\theta_3(\theta_2(\theta_1(\theta_0(\phi))))$, Post 1). Considering that Alice and Bob will probably associate further information with these types of players subconsciously, these two categories correspond to two new stereotype. The old stereotype is in fact discarded and not anymore part of Alice's and Bob's common ground (Post 2). Apart from that, there is no other information left unchallenged that would have also been grounded otherwise.

Apart from reaching mutual belief about ϕ or any of the incidentally communicated content during clarifications, agents also ground presuppositional and relational information as discussed before. Such information can only be induced by domain-dependent inference rules that we are not going to address here. Because these additional inferences are not properly established as mutual beliefs, however, the agents might come to different views about their common ground.

8.3.5 Grounding and Epistemic and Relational Goals

We discussed previously that people tend to adjust the information they communicate due to their common ground and epistemic and relational goals. We indicated that the interlocutors apply successive transformations to the communicated information in order to reach a common understanding. Obviously, these transformations cannot be arbitrary because then the goal of the speaker which information to transmit to further their task would not matter at all. We address this in the following, making the assumption that the joint activity description prescribes possibilities for modifying the information to be communicated such that this modification is still adequate for the purpose of the joint activity.

Let C the current intentional context of the interaction, i.e. the intentions and beliefs that the agent holds about this joint activity. Let Θ_ω the set of all possible transformations $\theta(\omega)$ on the proposition ω. We denote by $\Theta_w^C \subseteq \Theta_w$ the subset of Θ_w that contains those transformations that map ω to propositions ω' that are adequate to be substituted for ω in the context C of the current activity. An ω' is adequate if its transmission is as sufficient for the progress of the joint activity as ω itself. The proposition ω' could, for example, denote a concept or action that is more specific than ω.

Assume now that the agents can adopt intentions that epistemic or relational goals are achieved. When it comes down to which $\theta(\phi)$ is communicated initially and which further transformations from Θ_w^C are applied during the grounding process, these intentions will limit the agents' options. The effect of an utterance on relational goals relies on the content of common ground as discussed in Sect. 8.2. Consequently, a small Θ_w^C as probably applied during task-oriented dialogue will enforce accurate communication but limit possibilities for adjustment. Thus, grounding is likely to run into trouble and into requiring clarification in terms of further discussion and subordinate dialogue. In contrast, a larger Θ_w^C as for example applied during casual conversations will allow the agents to adjust their communication much better to epistemic goals, relational goals, and their common ground, thus facilitating the grounding of the proposition ω.

By adjusting the exchanged information, common ground can be extended in a way not originally intended by the speaker. Moreover, the speaker is able to build a different common ground about the same issue with different agents. For example, Alice suppresses the stereotype-inconsistent information that Gary drank wine when she talks to Maria. This information does not appear relevant to the task to work out how to improve Gary's reputation and might require additional discussion that could harm the relationship with Maria. However, any repetition of stereotype-consistent information in the conversation with her husband would just be boring and not add to their joint goal of having a conversation. Therefore, Alice transmits the novel information that Gary drinks wine. The joint goal of Alice and Stacy, in contrast, is to strengthen their relationship. The confirmation of mutually held stereotypes serves this purpose (Clark and Kashima 2007).

8.3.6 The Joint Activity and Grounding

If agent A needs to communicate some information ϕ to agent B, it will adopt the following goals that any of the $\theta(\phi)$ with $\theta \in \Theta_\phi^C$ becomes mutually believed:

$$\forall \theta \in \Theta_\phi^C [Goal_A(MB_{\{A,B\}}(\theta(\phi)))] \tag{8.3}$$

In the next step, the agent needs to commit to an intention that achieves one of these goals, which needs, however, to be compatible with the agent's existing intentions, including epistemic and relational goals. Let us denote by Σ the subset of Θ_ϕ^C where $\theta(\phi), \theta \in \Sigma$, is compatible with the agent's existing intentions. Then the agent will adopt the intention that it presents one of the $\theta(\phi)$ with $\theta \in \Sigma$ and that agent B accepts it:

$$\begin{aligned} &Int.Th_A\{Do[A, present(\theta(\phi))] \wedge Do[B, accept(\theta(\phi))] \\ &\quad \wedge Before(Do[A, present(\theta(\phi))], Do[B, accept(\theta(\phi))])\} \end{aligned} \tag{8.4}$$

The operator $Do(A,\alpha)$ is true if agent A does action α. By means-end reasoning, this intention-that entails another intention to issue the presentation: $Int.To_A\{present(\theta(\phi))\}$. We assume that when agent A is executing this presentation, agent B will recognize agent A's intention to communicate and thereby the agents will make the intention in Eq. 8.4 a joint intention, based on their commitment to the joint activity and the knowledge that agent A will not be able to achieve its individual intention until the communicated content is accepted (being responsive to each other):

$$Int.Th_{\{A,B\}}\{Do[A,present(\theta(\phi))] \wedge Do[B,accept(\theta(\phi))] \\ \wedge Before(Do[A,present(\theta(\phi))],Do(B,accept(\theta(\phi))])\} \quad (8.5)$$

Also agent B will then adopt an intention to perform its part of the activity. According to the properties of joint activities discussed in Sect. 8.3.2, each agent will be committed to the other party being able to perform its part. We denote this commitment by an intention-that. In particular, agent A will adopt the intention that agent B is eventually able to accept the presented information: $Int.Th_A\{Do[B,accept(\theta(\phi))]\}$. A successful performance of the joint intention will add $\theta(\phi)$—possibly transformed by clarifications—to the agents' common ground, as well as any unquestioned information exchanged during clarification subdialogues and any inferred presuppositional or relational information.

Given the properties of joint activity mentioned in Sect. 8.3.2, the agents will be mutually responsive to each other's intentions and they will support each other, also involving additional communication to maintain the consistency of their shared mental space. Thus if agent B cannot reconcile $\theta(\phi)$ with its beliefs and is unable to achieve its part of the joint activity, it will communicate this problem to agent A with a *req-clarify*. Agent B will thereby essentially open a sub-dialogue with a reason for its misunderstanding of A's original presentation and possibly a counter-proposal.

Now provided with an explanation why agent B cannot accept the information, and based on its commitment to the *accept*-action of agent B, agent A will react by further reasoning and answering to solve the misunderstanding. Therefore agent A needs to adopt further intentions in order to achieve the intention that agent B is able to perform the *accept*-action. The additional intentions will consist of a clarification of what has been said earlier, possibly presenting the communicated information in a different light. This corresponds to the grounding process as an alignment of mental models that we seek to represent.

8.3.7 Generalizing Context-Specific Common Ground

As discussed in Sect. 8.2, context-specific common ground can be generalized temporally, spatially, and socially. The first two options are implicitly represented in this model by agents changing their mental attitudes during the joint activity, thus making grounded information available to later activities of the same interlocutors. Two different parts of the common ground can be generalized socially, especially in relation to stereotypes: the subject- or "who"-part, and the target- or "what"-part. The former describes which agents are part of a common ground relationship. The latter describes information about the content.

Hence, to generalize the subject-part of their exchanged information socially, agents can agree that it is not only them who assume common ground of a proposition ϕ but also other agents in a group G (thus $CG_{\{A,B\}}(\phi)$ becomes $CG_{\{A,B\}\cup G}(\phi)$). To do so, one agent can provide information that implies that the interlocutors should adopt the belief that others also share their view. In effect, this can have an effect on the belief about who shares a certain stereotype and on the interlocutor's communal common ground.

To generalize the target-part of their exchanged information socially, agents can agree that their information $\phi(G)$ about a certain social target G (an individual or group) actually applies to a larger group $G \cup H$ (thus $CG_{\{A,B\}}(\phi(G))$ becomes $CG_{\{A,B\}}(\phi(G) \wedge \phi(H))$). This amounts to a modification of the actual information that is transmitted and can serve stereotype creation and change. For example, when Alice tells Stacy about Gary's being caught by police, she makes use of the phrase "the usual story" to indicate that the stereotype of football players as careless guys is actually shared on a communal level.

8.4 Discussion and Conclusions

This chapter has provided a tentative semi-formal analysis of the grounding model of cultural transmission, which emphasizes the micro-level dynamics of cultural transmission. The model postulates that cultural transmission happens during dialogue incidental to everyday joint activities, when interlocutors align their beliefs to a degree sufficient to carry out their joint activity. The description of this grounding process has relied on Clark's model of grounding in natural language use but has focused more on a higher level of discourse at which propositions are exchanged and their content is negotiated than on the production and interpretation of low-level utterances.

Towards a Computational Model We have built on intelligent agent research to explicate the link between agents' joint activities and the grounding process that is entailed by their task-oriented communication. Some computational models of collaborative discourse have been developed with either or both of these perspectives in mind. By far the most prominent computational model of Clark's grounding theory was proposed by Traum (1994), which is concerned with grounding at the utterance-level, not with the agreement about the exchanged propositional content we are interested in. Space does not permit even a brief account of computationally oriented research concerned with joint activities, but we provide a few pointers (Castelfranchi

1998; Cohen 1991; Dunin-Keplicz and Verbrugge 2002; Subramanian et al. 2006; Tambe 1997). In our subsequent work we have chosen to build on the *SharedPlans* framework of joint activity (Grosz and Kraus 1996).

A SharedPlan is basically a collection of intentions-that, intentions-to, and mutual beliefs that ensure consistent intentions between the agents participating in a joint activity (Grosz and Kraus 1996). Together with a set of axioms, these elements cause a SharedPlan to fulfill the requirements of joint activity identified by Bratman. Hence the SharedPlan formalism achieves the features we required of the kind of intentions-that described in Sect. 8.3.2. SharedPlans provide a clear and complete account of the collaborative planning and communication that agents might engage in during the course of a joint activity. Furthermore, agents that fulfill the requirements to reason about SharedPlans neatly correspond to our agent specification in Sect. 8.3.1. Therefore, we see potential for adopting SharedPlans as the representation of joint activities in our model. SharedPlans have been deployed together with a dialogue system to enable the cooperation between a system and its user for the achievement of a joint task (Rich et al. 2001). However, the particular characteristics of the grounding process we are interested in have not been considered. We also do not require a full-fledged dialogue system.

Contribution to Agent-Based Modeling of Cultural Dynamics One of the key features of agent-based models is their ability to explicitly represent and simulate micro-level interactions between individuals. Obviously, any such modeling depends on social scientific theories about these interactions. The grounding model of cultural transmission in particular offers a perspective on the role of micro-level interactions in cultural dynamics. The main contribution of this chapter is an explication of this model that can serve as a more precise framework for the specification of agent-based models of cultural dynamics. In that, we leverage the development of agent-based simulations of cultural dynamics that do take seriously the micro-level interactions between agents. This contributes to bridging the gap between micro- and macro-level of cultural dynamics.

Consider the transmission of stereotype-relevant information as an example to illustrate the prospects of this model. We have seen throughout this chapter that the transmission of stereotype-relevant information relies on the perception of interlocutors about the prevalence of stereotypes as well as the context of their interactions. While relational goals probably play a major role during interactions in public space, they are less likely to affect interactions between people that are already close. Given an implementation of this model, we could set up a network of agents in which members of the same communities are strongly linked but members of different communities are not. We would be able to simulate the transmission of stereotype-consistent and -inconsistent information within this population and observe the evolution of the stereotype's distribution at the macro-level and, in turn, its effect on micro-level interactions. Such a simulation would advance our understanding of stereotype formation, maintenance, and change in particular, as well as cultural dynamics in general.

Future Work The formalism is yet to be completed and we have also neglected the various internal processes of agents that determine which information to communicate. Having relied on a description in formal logics, we have neglected the fact that people are not omniscient and that their beliefs are not necessarily accessible at all times. Thus, a more realistic agent model constrained by bounded rationality would be desirable, but this would take us into much deeper waters. We have also suggested that agents would engage in the grounding process only to the extent that they deem necessary for their joint activity. We need to enable agents to make the decision what they consider necessary. We also need to equip them with the ability to estimate the cost of grounding a certain proposition, which, together with epistemic and relational goals, affects decision making.

References

Axelrod, R.: The dissemination of culture: A model with local convergence and global polarization. Journal of Conflict Resolution 41(2), 203–226 (1997)

Boyd, D., Richerson, P.J.: Culture and Evolutionary Process. Oxford University Press, Oxford, UK (1985)

Bratman, M.E.: Shared cooperative activity. The Philosophical Review 101, 327–341 (1992)

Castelfranchi, C.: Modelling social action for AI agents. Artificial Intelligence 103(1–2), 157–182 (1998)

Castellano, C., Fortunato, S., Loreto, V.: Statistical physics of social dynamics. Reviews of Modern Physics 81(2), 591–646 (2009)

Cavalli-Sforza, L.L., Feldman, M.W.: Cultural Transmission and Evolution. Princeton University Press, Princeton, NJ (1981)

Clark, A.E., Kashima, Y.: Stereotype consistent information helps people connect with others: Situated-functional account of stereotype communication. Journal of Personality and Social Psychology 93, 1028–1039 (2007)

Clark, H.H.: Using language. Cambridge University Press, Cambridge, UK (1996)

Cohen, P., Levesque, H.: Teamwork. Nous, Special Issue on Cognitive Science and AI 25(4), 487–512 (1991)

Dawkins, R.: The Selfish Gene. Oxford University Press, Oxford, UK (1976)

Dunin-Keplicz, B., Verbrugge, R.: Collective intentions. Fundamenta Informaticae 51(3), 271–295 (2002)

Gilbert, D.T., Krull, D.S., Malone, P.S.: Unbelieving the unbelievable: Some problems in the rejection of false information. Journal of Personality and Social Psychology 59, 601–613 (1990)

Gilbert, D.T., Tafarodi, R.W., Malone, P.S.: You can't not believe everything you read. Journal of Personality and Social Psychology 65, 221–233 (1993)

Gilbert, N.: Agent-based models. Sage Publications, London, UK (2008)

Goldstone, R.L., Janssen, M.A.: Computational models of collective behavior. Trends in Cognitive Sciences 9(9), 424–430 (2005)

Grice, H.P.: Logic and conversation. In: Cole, P., Morgan, J.L. (eds.) Syntax and semantics 3: Speech acts. Academic Press, New York, NY (1975)

Grosz, B.J., Kraus, S.: Collaborative plans for complex group action. Artificial Intelligence 86(2), 269–357 (1996)

Kashima, Y.: Conceptions of culture and person for psychology. Journal of Cross-Cultural Psychology 31, 14–32 (2000)

Kashima, Y., Klein, O., Clark, A.E.: Grounding: Sharing information in social interaction. In: Fiedler, K. (ed.) Social Communication, pp. 27–77. Psychology Press, Philadelphia, PA (2008)
Kashima, Y., Woolcock, J., Kashima, E.S.: Group impressions as dynamic configurations: The tensor product model of group impression formation and change. Psychological Review 107(4), 914–942 (2000)
Levinson, S.C.: Pragmatics. Cambridge University Press, Cambridge, UK (1983)
Lyons, A., Kashima, Y.: How are stereotypes maintained through communication? The influence of stereotype sharedness. Journal of Personality and Social Psychology 85(6), 989–1005 (2003)
Maxwell, D.T., Carley, K.M.: Principles for effectively representing heterogeneous populations in multi-agent simulations. In: Tolk, A., Jain, L.C. (eds.) Complex Systems in Knowledge-based Environments: Theory, Models and Applications, Studies in Computational Intelligence, vol. 168, pp. 199–228. Springer, Berlin (2009)
Queller, S., Smith, E.R.: Subtyping versus bookkeeping in stereotype learning and change: Connectionist simulations and empirical findings. Journal of Personality and Social Psychology 82(3), 300–313 (2002)
Rich, C., Sidner, C.L., Lesh, N.: Collagen: Applying collaborative discourse theory to human-computer interaction. AI magazine 22(4), 15–26 (2001)
Searle, J.R.: Collective intentions and actions. In: Cohen, P.R., Morgan, J., Pollack, M.E. (eds.) Intentions in Communication, chap. 19, pp. 401–415. MIT Press, Cambridge, MA (1990)
Smith, E.R., DeCoster, J.: Knowledge acquisition, accessibility, and use in person perception and stereotyping: Simulation with a recurrent connectionist network. Journal of Personality and Social Psychology 74(1), 21–35 (1998)
Smith, E.R., Collins, E.C.: Contextualizing person perception: Distributed social cognition. Psychological Review 116(2), 343–364 (2009)
Sperber, D., Wilson, D.: Relevance: Communication and Cognition. Blackwell, Oxford, UK, 2nd edn. (1995)
Subramanian, R.A., Kumar, S., Cohen, P.R.: Integrating joint intention theory, belief reasoning, and communicative action for generating team-oriented dialogue. In: Proceedings of the Twenty-first National Conference on Artificial Intelligence. pp. 1501–1506. AAAI Press (2006)
Tambe, M.: Agent architectures for flexible, practical teamwork. In: Proceedings of the National Conference on Artificial Intelligence. pp. 22–28 (1997)
Traum, D.R.: A computational theory of grounding in natural language conversation. Ph.D. thesis, University of Rochester (1994)
Tuomela, R.: Joint intention, we-mode and i-mode. Midwest Studies In Philosophy 30(1), 35–58 (2006)
Van Overwalle, F., Heylighen, F.: Talking Nets: A multiagent connectionist approach to communication and trust between individuals. Psychological Review 113(3), 606–627 (2006)
Van Rooy, D.: Modeling multidirectional, dynamic social influences in social networks. In: Anderssen, R.S., Braddock, R.D., Newham, L.T.H. (eds.) MODSIM09 International Congress on Modelling and Simulation (2009)
Van Rooy, D., Van Overwalle, F., Vanhoomissen, T., Labiouse, C., French, R.: A recurrent connectionist model of group biases. Psychological Review 110(3), 536–563 (2003)
Wooldridge, M.: An Introduction to MultiAgent Systems. John Wiley & Sons, Hoboken, NJ, 2nd edn. (2009)

Chapter 9
Matching and Mismatching Social Contexts

Bruce Edmonds

9.1 Introduction

Context is everywhere in the human social and cognitive spheres but it is often implicit and unnoticed. When trying to understand and model the social and cognitive realms it has become increasingly clear that context is a crucial factor. In particular this chapter will argue that is it vital if one is going to map how different cultures can relate to each other.

This chapter starts by discussing context in general to distinguish some of the many meanings of the word "context". It then goes on to briefly discuss the pervasiveness of context-dependency in human cognition and how social contexts acquire their distinct identity, becoming entrenched within a culture. The next sections, which analyse how cultures might map onto each other in terms of their different social contexts, are the core of the chapter. These sections go into some detail about the various possible cases of match/mismatch both in terms of the identification of a social context as well as the assumptions, norms, habits etc. that are associated with these contexts. The chapter ends with the implications of this in terms of integrating cultures and studying such integration using agent-based simulation.

9.2 About Context

The word "context" is used in many different senses and has many different analyses (Hayes 1995). It is somewhat of a "dustbin" concept, in that if a theory or idea does not work the reason may be assigned to "the context". Thus for many (e.g. linguists)

B. Edmonds (✉)
Centre for Policy Modelling, Manchester Metropolitan University, All Saints Campus, Oxford Road, Manchester, M15 6BH, UK
e-mail: bruce@edmonds.name

V. Dignum and F. Dignum (eds.), *Perspectives on Culture and Agent-based Simulations*, Studies in the Philosophy of Sociality 3, DOI 10.1007/978-3-319-01952-9_9,

context is a subject that is generally avoided due to its difficulty. I cannot touch on all the approaches to and models of context in the literature, but will give a brief introduction to context in general, including some major conceptions of it, and a few of the issues surrounding it. This will, hopefully, clear the ground for the main suggestions of this chapter and avoid *some* of the possible confusions.

9.2.1 Situational Context

The situation context is the actual situation where some events or other described phenomena take place. This could include the time and location, but could include all that is the case about that situation, including: who was there, the knowledge of those people, the history of the place and the objects present. In this sense the context is indefinitely extensive, it could include *all* the circumstances in which an event occurs.

Such a context may be able to be specified adequately (if rather uninformatively) by giving the time and place of the event,[1] but the relevant details might not be effectively retrievable from this. For example, the fact "I was reminiscing about our summer holiday" might well not be detectable from the time and place except by the person doing the reminiscing. Thus when talking about the situational context it is common to abstract from this to that which is relevant about that context, or what might be commonly understood. Thus the phrase "the context" (as in the question "what was the context?") may mean "those factors that are relevant to understand this particular occurrence" even though it may refer to the original situational. Thus to understand what someone is saying to you, you might ask "what was the context?" and get a description of the circumstances, e.g. "I was on the train with my mother".

9.2.2 Linguistic Context

Whilst the situational context could include anything, at least in theory, the linguistic context is composed of the words that surround an utterance or phrase. This typically indicates the words that precede or frame the target of understanding, but could also include common knowledge that could be reasonably be expected to be known by the listener/reader, e.g. elements of the relevant culture. Sometimes this is taken to be the same as all that which is necessary to understand some natural language.

[1] As is essentially the approach in (Barwise and Perry 1983).

Historically this has been what one appeals to if there seems to be no detectable foreground features to explain some particular meaning. However more recently more positive attention has be focused on context in linguistics. For example, Peter Gardenförs has said ("pragmatics" being a term that includes contextual considerations in linguistics):

> Action is primary, pragmatics consists of the rules for linguistic actions, semantics is conventionalised pragmatics and syntax adds markers to help disambiguation (when context does not suffice). (1997)

Clearly the linguistic context could refer to almost any of the language or culture that surrounds an utterance, and hence is not something that can be captured in its entirety. Often context is thought of as linguistic context because the interactions that are being considered consist of linguistic communication.

9.2.3 Cognitive Context

Clearly many aspects of human cognition are context-dependent, including: visual perception, choice making, memory, reason and emotion (Kokinov and Grinberg 2001). What seems to occur is that the human brain categorises kinds of situation which it is able to later recognise, largely without conscious effort. A lot of recall, learning and inference is with respect to these recognised kinds of situation. This abstraction of a situation in the brain – the recognised kind of a situation – is the *cognitive context*. It is the cognitive correlate of the situational or linguistic context, abstracting and limited by what is relevant. Such cognitive contexts could be identified using a description of the kind of situational context that invokes them or else by the set of all the knowledge, norms, expectations, habits etc. that are immediately accessible once recognised.

It is essential that the circumstances under which different cognitive contexts apply can be effectively and reliably recognised but this does not mean that they have to be consciously recognisable as distinct contexts and labelled. They may be unconsciously recognised by all the members of a community but never named; maybe they their features are distinctive and consciously recognisable but too complex and fuzzy to be completely specified.

That we flexibly learn to recognise contexts and what is appropriate to them that allows for the culturally-specific development and entrenchment of social contexts.

9.2.4 Social Context

Many of the cognitive contexts we have learnt seem to correspond to recognisable kinds of social situation. Examples include: greeting, lecturing, and a political discussion. Once established these seem to be self-perpetuating: habits, conventions, norms, terms etc. are developed by people who recognise the context, but the fact that there are specific habits etc. that pertain to this makes them more recognisable. Thus social contexts can be co-constructed over time and passed-on to others. Individuals co-learn the cognitive correlates along side others so that these correspond to the kind of social situation (similar to how we learn to use the same word for the same colour).

When people are asked to describe the context, they will often do it in social terms. Thus it is that the social context, although it is a special case of situational context is closely linked to the synchronised cognitive context that participants have learnt to associate with situation, because it is often the social aspects that are important in terms of communication and understanding. It is, of course, social contexts that we will be primarily concerned with in this chapter, although these must largely correspond to the cognitive contexts in the minds of the society's.

9.2.5 Identifying and Talking About Context

One of the difficulties in discussing context is that they may well not (a) be accessible to us (b) identifiable even if they are accessible or even (c) definable in precise terms even if we can identify them. This is due to the complex and largely unconscious way in which context is recognized. Rather we often have to try and deduce what the relavant contexts are by introspection and other observation.

Despite this, we often *talk* about contexts as if they were discrete "things", however it needs to be understood that *for our conscious selves* this may not be the case. Thus "the" context is an abstraction of the aspects of those background features that define it, whether or not this is a meaningful or reifiable entity for us. To simplify the discussion I will generally talk about contexts in the sections below *as if* they are well defined identifiable entities, but the caveats just mentioned need to be always taken into account. This difficulty means that *the* context for any situation is often not made explicit or represented – those involved may well not be aware of the cognitive context they are assuming.

Thus although we may not be able to describe or specify cognitive contexts in general due to their fuzzy, complex and inexact nature, social contexts are more identifiable due to their institution into the fabric of our society and the fact that they need to be readily recognizable by all actors.

To aid the discussion to follow I will use some shorthand terminology. *The context* means the recognized kind of situation where specific rules, norms, language, dress, habits etc. may hold. The *contents* of a context are those specific rules, norms etc. that are associated with that context that will be widely known by the social actors participating in them. The analogy is of a set of containers of these specific items of knowledge that are taken off the shelf and accessed depending on the recognized kind of situation.

9.3 The Context-Dependency of Human Behaviour

9.3.1 The Pervasiveness of Context-Dependency in Human Cognition

People behave differently in different situations – the rules, norms, expectations and decisions that people apply in "a lecture" will be very different from those they apply at "a celebration". I would claim that this is not a coincidence, nor only a surface difference masking underlying universal patterns of behaviour – but that context-dependency is fundamental to the way humans deal with the world. There are several reasons why this claim is a strong one. *Firstly*, there is the simple observation that the same people behave very differently in different social contexts and that different people behave in similar ways in the same social context. *Secondly*, there is a lot of evidence that human cognition is context-dependent in many respects. Kokinov and Grinberg (2001) list some, including: visual perception, choice making, memory, reasoning and emotion, preferences, and language comprehension. Context-dependency seems to be hard-wired in our cognition, and it would be very surprising if this did not result in a context-dependency in terms of behaviour. *Thirdly*, one can see why it might have evolved in our species. If, as seems likely, a significant part of the evolutionary advantage that our brains provide us is in our ability to organise and adapt in social groups – as suggested in the "Social Intelligence Hypothesis" (Kummer et al. 1997) – then the ability to learn to behave in a highly context-dependent manner can be explained in terms of the significant advantage that would result from *groups* being able to develop different shared norms, habits and protocols to suit different tasks and situations.[2] So, for example, how to behave on a fishing expedition can be developed to suit the conditions and technologies available for that activity, but the patterns of the whole group would change quickly and simultaneously if a conflict with a competing group arose, or a storm was coming.

Human cognition seems to involve the combination of rich, unconscious, fast and vague context recognition with relatively simple, conscious, slow and precise reasoning and learning relative to the context (Edmonds 1999a). Dividing the world into similar kinds of situation and dealing with it on that basis makes the conscious reasoning, learning, and decision making feasible (Greiner et al. 2001). The flexibility of this combination or rich Machine Learning kind of mechanism with slow but specific Artificial Intelligence style reasoning and adaptive mechanisms seems very effective and powerful for the particular environment and social existence that we have (Edmonds and Norling 2007).

[2]It might be that just as language might have co-evolved with the brain (Deacon 1998) that this ability to coordinate via a shared social context might have co-evolved with our cognitive abilities to deal with context.

One of the features of this combination is that the recognition of the kind of context is usually unconscious (unless, for example, we misidentify the appropriate context). Thus our brain automatically (and apparently seamlessly) is providing us with the knowledge, expectations, habits, vocabulary to deal with the kind of situation we are facing, without apparent cognitive effort. This rich and unconscious context recognition can make it hard to indentify or even talk about context, which is perhaps why it has not had the attention that it deserves.

Also, it must be said, that there has been a bias against context-dependent understandings of social and psychological phenomena on the grounds that it is not "scientific". It is true that universal models and understanding is preferable if they are possible, but there is no reason to suppose[3] that the world has been so conveniently arranged for us in this respect (Edmonds 2012).

Regardless of our ultimate philosophical views on the existence of universal underlying mechanisms, in practice it is sensible to understand and model human behaviour as context-dependent. This is particularly true in the social and cultural sphere where reductions to putative universal underlying mechanisms are currently no more than a theoretical commitment.

9.3.2 The Development and Entrenchment of Social Contexts

Whilst general types of environmental situation may well be identified and learnt by people (paths, clearings etc.) it is in the social sphere that context is delineated in the most obvious manner. This is because situations that are recognised as a kind of context become entrenched as the result of social processes. Thus if a situation occurs, such as a lecture, then it may be recognised by others too. Over time particular rules, norms, ways of behaving and language might be invented to suit that kind of situation. The more particular things pertain to it, the more clearly it is recognisable. The more clearly it is delineated the more it is likely that things will be invented or adapted for that kind of situation – the context becomes socially entrenched via the co-development of its identity and specific content.

For example, consider the *lecture*. It is likely that early lectures were held outside or simply in people's houses. They might well have been much more fluid than the lectures we are used to, with people coming and going during it, and more of a dialogue, or even barracking, during the lecture. The lecture may not even have been distinguished from other kinds of teaching or discussion. However over the years the lecture has developed into a sharply defined institution. We build special rooms to hold lectures in. People are trained from an early age how to behave in a lecture, that it is not allowed to disrupt a lecture, that one can expect the lecturer to have some expertise in what they are talking about, that limited amounts of questions will be

[3]Other than sheer optimism.

allowed to be put but not free-form dialogue. It is now trivial to recognise a lecture and we all know how to behave and what to do in them.

All cultures, through necessity or invention develop a series of kinds of situation that have different purposes or provenances and which are correctly recognised by the members of the culture. That is, it is not only the expectations, habits, language, actions, norms, dress etc. that will have developed differently in different cultures but also *when* different sets of these pertain. These kinds of social context structure the different cultures, in ways that its members recognise. In other words, social context has a crucial role in the social embedding of individual action (Granovetter 1985).

This is not to say that *all* culture reduces to considerations of context, but just that context-dependency so permeates human social action that it is impossible to ignore. Within a single culture, among acculturated members, knowledge about its social contexts will be common, so there is no need to make the contexts explicit. This and the fact that context is omnipresent and unconscious means that little is written about the relationship of context and culture. When considering how cultures might relate to each other, one cannot leave context out of the picture. The "contents" of social contexts will also differ profoundly from those of other cultures, *even* within closely identified contexts. However this chapter is focussing on the *extra* dimension that emerges when each culture has different contexts, *when* different cultural elements are relevant, as well as that of *how* these elements might differ or relate.

9.4 Implications for How Different Cultures Map onto Each Other

The principle point of this chapter is to point out that the existence and identity of the entrenched social contexts in cultures will have a huge impact as to how those cultures relate to each other. That is to say, the *relative structure* of cultures matter.

Sometimes the social contexts that have developed within different cultures will roughly coincide. That is to say the broad identification, function and style of a social context in one culture will be identifiable in another, even if the content and scope differ somewhat. Thus a service of religious worship may vary greatly from culture to culture in terms of what is expected of people within that service, when it happens, its place in relation to other aspects of society etc. etc. but is still broadly recognisable as such. Other examples of this include a court of law or a wedding.

Why these are identified as being of the same kind, or indeed why they are even identifi*able* as the same kind is complex and sometimes mysterious. It may be that the social institutions have travelled across the world with the spread of technology, people, trade, empires or religion. It may be that the commonality can be traced to a need or function that is the same for people everywhere, for example a funeral or cooking. It may even be that situations with different roots come to be identified as belonging to the same category, for example a musical performance. Of course, it may be that some social contexts that are identified as the same have, in fact, little in common but some relatively unimportant surface features. A meal might involve some eating by definition, but the meal of a solitary diner experiencing a variety of flavours in a *haut cuisine* restaurant might have nothing much to do with the ceremonial eating of unleavened bread at the Jewish Passover festival.

In other cases the social context in one culture may not have anything that corresponds in the other. So the social context of a commuter train may not have much in common with travelling in a nomadic community. Sometimes a new social context appears that does not spread to other cultures, or has not spread yet. There may be no need for a babysitting circle in places where childcare needs are dealt with purely within an extended family. What is acceptable in one culture might not be acceptable in another, so a "rave" (where large numbers of young people gather to dance to hypnotic music with an accompanied use of drugs and alcohol) would simply not be tolerated in many countries.

Of course, deciding whether or not a particular social context in one culture corresponds to another in a different culture is highly problematic. Social phenomena are horrendously complex, changing and subjective. It may even be that, ultimately, it is impossible to come up with precise identifications that can stand up to rigorous questioning. However this is not the point. However problematic the identification of social context across cultures is, the fact is that people *do* identify some social contexts as similar across cultures and this effects their judgements and reactions to elements of different cultures. How these social contexts are perceived, and the social consensus on this matter *does* have a definite consequences for how and when cultures can integrate.

The central point of this chapter is that integrating an element of one culture with another will be very different in cases where the social contexts are widely considered to be of the same kind to those cases where the social context of one of the cultures is not recognised as having an equivalent in the other. That is to say that the structure of social contexts matters, as well as the "content" of those contexts. The reason for this is that the *scopes* of social contexts (*when* they are considered to occur) are very difficult to change, often deeply embedded within a culture, and to a considerable extent unconsciously assumed.

This analysis implies that cultural integration will be fundamentally affected by the structuring of social contexts in each culture. The case where the contexts largely overlap (e.g. an academic lecture) will be very different from when they do not (e.g. some religious contexts). This, in turn, implies that there will be some very different kinds of cultural integration corresponding to these different cases.

9.4.1 Different Kinds of Integration that May Occur

Let us consider some of the possible cases of correspondence, or lack of it that might occur when different cultures have to co-exist and hence encounter each other. Table 9.1 lists four basic cases of match/mismatch: where a social context in culture

Table 9.1 Four cases of match/mismatch

	The scope of social context in culture A **does not** correspond to anything in culture B	The scope of social context in culture A **does** correspond to a social context in culture B
The "content" of social context A is **roughly compatible** with culture B	***Case 1*** Compatible new	***Case 2*** Roughly compatible
The "content" of social context A **significantly clashes** with culture B	***Case 3*** Clashing new	***Case 4*** Internal clash

A does/doesn't have a corresponding social context in culture B, and where the "content" of the social context in culture A is compatible/clashes with culture B. Of course the author recognises this is a simplistic categorisation but these differences come from pragmatic social considerations and facilitates discussion of each case. I have given each case a simple label to make the exposition clearer.

9.4.1.1 Case 1: Compatible New

In this case there is a social context in culture A that does not correspond to anything much in culture B, but the content of that social context is largely compatible with the norms, ethics, habits etc. of culture B. Thus a new pastime that is transplanted with the movement of people to a new culture can simply add to the menu of choices in the receiving culture, and recognised as such. Indeed it may be that it becomes a cherished context within the receiving culture and, over time, becomes embedded within that culture. This case is largely unproblematic, since there is no confusion that the context is new, nor any competition as to the detailed identification of that kind of situation. There may have been no such thing as "homework" (a specific piece of work for the pupil to do out of school and handed in to be marked) in some cultures, but as a potentially useful addition to the life of children can be simply added-on if it does not conflict with other patterns or duties.

9.4.1.2 Case 2: Roughly Compatible

In this case there is a social context in culture A that many people would identify as essentially the same (or indeed "the same") as a social context in culture B, and the nature of what happens in the context in culture A is roughly compatible with that in culture B (it is almost impossible for it to be identical, even with an institution trying to standardise them). This case is problematic in so far as people might simply assume the content and identification of the situations to be completely the same when there are, in fact, small differences. Thus instead of dealing with a situation using well-entrenched habits associated and triggered by the context, some conscious thought and adaption might be necessary. Thus if one goes to a religious service in another country, even one within the same global institution, then there might be differences in norms about lateness, talking during the event, style of dress, etc. An unthinking reaction might lead to a negative reaction to these differences (e.g. "they had no respect, talking and chatting all through it!").

In this case bringing the differences to the foreground, pointing them out and the different reasons and roots of the differences, i.e. explicitly educating people about the differences can help ease any dissonance that might have occurred. The trouble is that the habits and assumptions associated with particular social contexts are largely automatic and unconscious otherwise. Thus, in both US and UK cultures there is a well-recognised social context of "greeting and getting to know a person" within such events like parties, meetings etc. However (to generalise broadly) it is largely the norm within US culture to tell the other about oneself as a way of opening up the conversation and in the UK it is the norm to ask the other about themselves for the same purpose. This can lead to the case where after a first meeting between a US and UK citizen they come away with negative impressions of each other ("the US are always bragging about themselves and never once asked me what I do" and "the UK people are so snobby and close he did not tell me anything about himself but kept me at a distance" being stereotypical reactions in this case). Sometimes simply pointing out the differences can be enough to sort these misunderstandings out, in other cases habits are so ingrained or beloved that people are unwilling to adjust possibly leading to minor mutual irritation.

9.4.1.3 Case 3: Clashing New

In this case, there is a social context in culture A that does not correspond to anything much in culture B, i.e. it is "new", but the content of that social context is either incompatible with the norms etc. of culture B or it is perceived as being problematic from the point of view of Culture B. The "intruding" social context is not identified with social contexts in the receiving culture by many of those in that culture. Of course such an intrusion may be a matter of perception and not, ultimately, a matter of practical, legal or moral incompatibility. Thus there may be a neutral or positive reaction from people who identify a Mosque with a Church, and Islamic prayer and worship with Christian prayer and worship to the plan to open an Islamic Centre near the site of the 9/11 terrorist attack, but a negative one from those who see Islam as basically alien to them – an intrusion into their society. Whether the incompatibility is real or perceived, the conflict it can trigger might be very evident in the form of peoples' actions and rhetoric. Without making a judgement there are several possible outcomes to such a situation (in general).

The receiving culture might decide to ban or discourage the new kinds of social situation from taking place. This might or might not be successful, depending on the lengths to which they are prepared to go, the level of conflict with the sending culture that they are willing to tolerate, and the importance and embedding of the social context within the sending culture. There are cases where this has been largely successful (e.g. monogamy even for immigrants from countries where polygamy is practiced and legal), and cases where it has not worked (e.g. the past attempted suppression of churches in China).

After a time it might be that the new context is accepted in the receiving culture. As new generations grow up with the "problematic" social contexts being part of their social environment, it may not seem so threatening. Of course, this depends on the incompatibilities not being fundamental, so this acceptance will occur if either: the perceived incompatibilities become accepted over time, or the incompatible elements are adjusted by the incoming culture so that it is acceptable. Such a process of acceptance and/or adjustment can be facilitated with the involvement of the receiving culture's people in the introduced context (either directly though participation or indirectly via social contact) both in helping correct misperceptions but also in the introduction of the receiving culture's values to those in the incoming culture.

9.4.1.4 Case 4: Internal Clash

In this case the social context of the incoming culture is identified with that in the receiving culture, but there are incompatibilities in terms of what happens within that social context. For example, "waiting for a bus" is a social context that is recognised widely throughout many cultures, but the norms of how one behaves when the bus arrives might differ. There is an obvious and direct conflict between the norms of queuing (the people enter the bus in the order they arrived at the waiting place) and walking on without regard to order as quickly as possible. The former might well resent the latter as "pushing in" illegitimately, and the later might be frustrated at what they perceive as the "unnecessary" formality concerning a simple action of entering a bus, especially in cases where there is room for all. Both views may react on the basis of deeply entrenched habit and norms, and resent the other pattern of behaviour, often attributing onto the others bad motivations and character.

In some of these cases there is no easy resolution, but that one or other pattern of behaviour will win out. It may be that in the long run social influence determines the outcome, either newcomers are persuaded to adopt the norm of queuing in the UK, or it may be that this norm falls into disuse. Sometimes these clashes are decided by enforcement, with reference to "fundamental" rights and duties of a society that incomers or inhabitants must abide with. In France it was decided that wearing the *hijab* in public was illegal – thus in this case that the incomers were forced to adapt to the existing norms. In the US it may be that the principles of freedom of religion and expression would make such a law impossible, and it be the duty of the receiving culture to be tolerant.

However another response to this case (Internal Clash), is to encourage the differentiation of contexts, so that the contexts with respect to culture A and culture B come to be considered as separate kinds of situation and thus avoid the conflict due to the expectations of people that they are the same. Perhaps an example of this is the variations of the academic lecture that may be found in some of the more traditional Islamic world, where the lecture hall is arranged so that male and female students are screened from each other.

9.4.1.5 Case 5: New Contexts

A case that is not covered in the above classification is when a new social context is created that is separate from those in either culture. Such a context might be created to ease the interaction of different cultures, since neither culture will have engrained expectations of such a context. This might reduce the misunderstandings that might arise, since all participants are aware they are not in any of their "home" contexts. An example of this might be the international business meeting, which has gradually evolved to be distinct from a normal business meeting in any one country.[4]

9.4.2 The Implications for Attempts to Promote Social Integration

If it is indeed the case that the social contexts, along with their associated habits and expectations are relatively difficult to change (once entrenched), then this has consequences for what might be effective at promoting cultural integration or, conversely, avoiding cultural conflict. What is likely to work will be dependent on which of the cases above one has. Case 1 is largely unproblematic. In Case 2, education (explaining the differences) and making the unconscious assumptions explicit, bringing them into the open might be effective. In Case 3 there might well be no adjustment possible and the receiving culture simply has to decide whether it will tolerate the "intrusion". In Case 4 the only thing to do is adjudicate as to what can occur and what the fundamental rights are, hoping that time makes the alien familiar. Examples of Case 5 should simply be encouraged as the easiest medium-term approach to establishing working interaction and dialogue.

Of course, these prescriptions are simplistic and imprecise, more of a starting point for investigation rather than its conclusion. The dynamics of perception, context and interaction can easily make them otiose. Also the fractal nature of social contexts (sub-contexts within contexts etc.) can make a useful analysis of how contexts might correspond complicated. Finally what a social context *is* depends crucially on how they are perceived, and so changes of perception might result in apparently radical and quick changes in the scope of social contexts. However, it is also equally clear that a generic approach to considering how cultures relate and/or integrate which does not take their contextual structure into account might be woefully inadequate.

---

[4]The fact that such a context derived from that in a particular context does not prevent it developing into a new and separate context.

9.5 Simulating the Integration of Cultures

9.5.1 Representing Both Cognitive and Social Aspects of Social Context

Clearly there is a lot that is not known about: how social contexts and social norms interact, how they develop and fall into disuse, how people recognise social contexts, how group identity and signals and social contexts interact... etc. etc. To start to understand and unpick these complex and complicated interrelated factors we need to simulate them – since there will be a severe limit to how much one can keep track of these implications informally. The micro–macro link here is important, in other words it is essential to understand *how* the abilities, biases and intentions of individuals determine and are determined by the higher-level social.

If the above analysis is at all correct then (a) the awareness, identification and representation of social context is essential to fully understand how cultures might integrate and (b) the dynamics of cultures, with their constituent individuals, social contexts, norms, habits, assumptions and interactions could be highly complex. Together these indicate that in order to get a fuller picture, agent-based simulation is the most appropriate tool (Edmonds 2010b). This is because this technique can (a) represent some of both the cognitive and social aspects involved in social context and (b) track some of the complex interactions between social context, perceptions, habits, norms and actions of the individuals concerned.

However simulations that take seriously *both* the cognitive and social complexity in terms of what is represented is rare. It has tended to be that in many social simulations the KISS principle[5] rules as far as the cognitive model of the agents is concerned, concentrating on how complexity can emerge from the interaction of many relatively simple individuals. This is the approach exemplified in (Axtell and Epstein 1996). Clearly one can use such simulations to discover possible ways in which social complexity could occur, but this does not tell us how complexity in observed human societies occurs.[6] There *is* a community that takes the representation and simulation of human cognition seriously – the cognitive modelling community. This community does seek to represent in detailed simulations how we think. However it is very much from the individual point of view – understanding how an individual thinks. The social situatedness of human cognition is rarely touched upon here, and thus explorations of how the social embeddedness of human social artefacts, such as social context emerge, are maintained etc. are not possible.

---

[5]"Keep It Simple Stupid!", the engineering principle that one should only introduce complexity after simpler approaches have failed – the opposite is "KIDS" (Edmonds and Moss 2005).

[6]Of course one can make the heroic assumption that the nature of human cognition does not matter when it comes to the social layer (e.g. Ye and Carley 1995; Gilbert 2006) argues that one does not *always* have to accurately model cognition in social simulation. However (Edmonds and Moss 2001) shows that the cognitive model can be crucial.

However there are some projects that are starting to include representations of both cognitive and social complexity in their simulations. An example (though not about social context) is the EMIL project (Conte et al. 2010) that sought to simulate the twin cognitive and social aspects of social norms, allowing the exploration of norm emergence in terms of both mental perceptions of obligations and the social patterns that co-developed with these. A similar project is needed to start detangling the co-dependence of individual learning and perceptions of social context and their institution and embedding within the practices and artefacts in society.

9.5.2 Some Existing Simulation Work that Points in These Directions

It is notable that very few social simulations represent any of the processes for dealing with such context-dependency. That is to say, the agents in social simulations tend to be endowed with cognitive processes which are not sensitive to, recognise or use context. If the situation in which the agents are being represented can be considered as a single entity, so that all interaction can be considered as taking place within a single agent, then this is reasonable since one then does not need context.

However many simulations aspire to be a more general theory of social interaction. In this case, one has to assume that either the simulation is to be taken only as an analogy or that the simulator thinks that people's behaviour, norms etc. will be so similar between social contexts that including mechanisms of context recognition, dependency, etc. are unnecessary (Edmonds 2010a).

In the former case where the simulation is used only as an analogy, then this is valid because humans are experts at applying analogy in a context-dependent manner, adjusting its assumptions and form to be appropriate to its domain of application.

In the later case, where an essentially context-independent algorithm is used to represent a highly context-dependent process must, at least, be the legitimate target for doubt. Whilst the psychological realism that is necessary in a social simulation does depend upon the purpose of the simulation and the level of aggregation (Gilbert 2006), it is certainly not the case that the results of a simulation can be assumed to be robust against changes in the cognitive model being used (Edmonds and Moss 2001).

There are not many simulations which represent aspects of context-dependency in their agents, but there are a few: (Edmonds 1998) used a cognitive learning model specifically because it included some aspects of context-dependency; (Schlosser et al. 2005) argue that reputation is context-dependent, (Edmonds and Norling 2007) looks at the difference that context-dependent learning and reasoning can make in an artificial stock market, (Andrighetto et al. 2008) shows that learning context-dependent norms is different from a generic adaption mechanism, and

(Tykhonov et al. 2008) argue that the definitions of trust mean that trust is also context-dependent. (Alam et al. 2010) present a model of the exchange and family structure within a Mexican village based on the cognitive model of choice called "endorsements" (Cohen and Grinberg 1983). They justify the choice of this particular mechanism by the way it can result in context-sensitive choice.

These show that, at least in some cases, that context-sensitive cognition can make a difference. The fact that it can make a difference is not very surprising given the important role it plays in human cognition and society.

There are approaches to including cognitive context within the learning and decision-making of agents. (Andrighetto et al. 2008) use an approach based on social norms, whereby some of the habits and knowledge of agents are dependent upon the social context, in the sense of which group they are part of. (Edmonds 2001) suggests a particular algorithm and approach to learning appropriate cognitive context (which is discussed further below). This showed that the knowledge could be learnt and recognised in a way so that distinct contexts emerge, but it did not show that agents co-learnt the same social contexts, due to the anti-coordination motivation provided by the artificial stock market environment the agents inhabited.

It must be said that cognitive contexts that implement cognitive dependency are thin on the ground. This indicates that more work, both foundational and applied is needed if social simulations that can start to represent social context are to be become useful. However the next section looks at some possible ways forward in this respect.

9.5.3 Towards Implementing Simulations that Incorporate Context

Broadly there are three feasible approaches for incorporating context-dependency within a simulation. These are as follows.

1. *Within a Single Context*. That is focus on one, identifiable social context and only model agents acting within this.
2. *Within a Fixed Set of Known Contexts*. Identify *N* social contexts and model the behaviour and interaction within each of these, along with the agents' recognition of when a shift in context is appropriate.
3. *Learn to Identify Context with the Behaviours*. Let the agents in a simulation learn what the appropriate contexts and context-specific knowledge are.

The first of these is "modelling as normal". A context is identified (or assumed) and the behaviours of agents are simulated in the normal way. The behaviour of agents is based upon a fairly simple cognitive model, incorporating those elements deemed to be relevant for the behaviours in that context. All of the interactions are presumed to occur, *essentially*, within that context. As long as the context within which the model is valid is reliably recognised by all concerned, either explicitly or implicitly, then the context can be taken for granted, as long as extra-contextual influences do not significantly affect the outcomes (Edmonds 2009).

The second case requires the relevant contexts to be identified, with an essentially separate behavioural model for each context (though there may well be commonalities between them). In addition the rules for when the actors switch contexts need to be modelled. This makes for a lot of work, however by restricting each behavioural to a particular context, it may be that the context-specific models are fairly mundane and simple. This is, after all, what Herbert Simon observed in his observations of administrative behaviour, what he called "procedural rationality".

The human being striving for rationality and restricted within the limits of his knowledge has developed some working procedures... These procedures consist in assuming that he can isolate from the rest of the world a closed system containing a limited number of variables and a limited range of consequences. (Simon 1976)

Here standard elicitation and ethnographic techniques from the social sciences could be used to lay bare the procedures that people have developed, maybe starting from first-person accounts of what they do normally and some variations from this. However it is necessary to "prime" interviewees with each intended context – maybe presenting each of them via vivid scenarios – to get the full information. Although a number of techniques exist to help analyse and "make sense" of such narrative data, there still lacks a complete methodology to bridge the gap from such data to agent rules in a simulation. Currently this tends to be done informally by the programmer.

The third of the options is to equip the agents in a simulation to identify and learn both the relevant contexts and the behavioural rules themselves. In other words, nothing less than allowing the agents to acculturate within the situation (including the society of other agents) that they find themselves in. There are several major difficulties with this approach. *Firstly*, there are no mature methods for implementing such a cognitive mechanism, athough I sketch the one I have explored below (Edmonds and Norling 2007). *Secondly*, the process of acculturation is slow, needing a considerable amount of time so that the agents can learn and adapt to their environment. *Thirdly*, this requires the specification of an environment for them to adapt to. Since social environments are what we are primarily concerned with here, we have the problem of initialisation of a social environment. To produce a social environment one needs socially adapted agents, but one needs a social environment for these agents to adapt to. This is the deep problem of social embedding (Granovetter 1985; Edmonds 1999b). One can gradually co-evolve the agents to be each others' social environment, but then one cannot easily direct this to be the same as those observed. However, this might be used to reveal some general patterns and issues concerning cultural integration.

I do not have space to go into such an architecture, but only briefly illustrate it, see (Edmonds and Norling 2007) for details. The architecture is illustrated in Fig. 9.1 below. The memory (CDM) is spread over a space, so that given position in this space (which is the cognitive correlate of the context) beliefs are retrieved and laid down near this position – this neighbourhood is the cognitive context. There are three other components which all access the memory: a machine learning module that guesses the cognitive context from a rich selection of perceptions (the

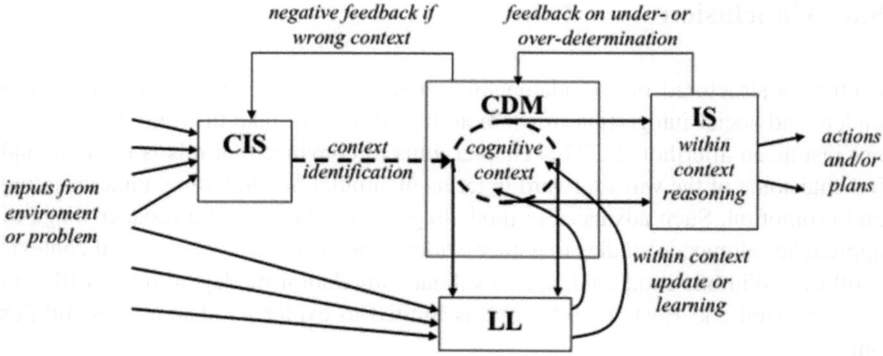

Fig. 9.1 How the context-identification system (CIS), the context-dependent memory (CDM), the local learning algorithm (LL), and Inference system (IS) might work together

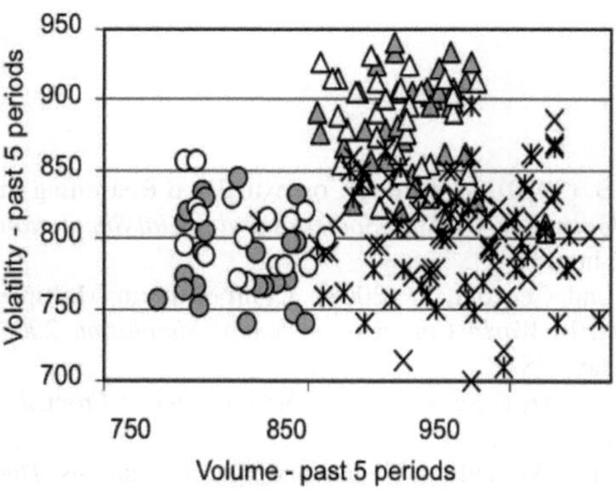

model-256	priceLastWeek [stock-4]
model-274	priceLastWeek [stock-5]
model-271	doneByLast [normTrader-5] [stock-4]
model-273	IDidLastTime [stock-2]
model-276	IDidLastTime [stock-5]
model-399	minus [divide [priceLastWeek [stock-2]] [priceLastWeek [stock-5]]] [times [priceLastWeek [stock-4]] [priceNow [stock-5]]]

Fig. 9.2 The emergent clustering of mental models within a trading agent's context-dependent memory, (*left*) the position of the most frequent mental (*right*) the models

CIS above); a belief revision module (the LL above) that updates the set of beliefs within the cognitive context; and a decision making/reasoning module that takes the beliefs within the cognitive context and makes decisions or plans which determine action.

Figure 9.2 shows the emergent clustering of beliefs within an agent in a simulation of a stock market, which uses the above architecture. This is the set of models to predict the future prices of a particular stock by a particular agent at a particular point in time. The memory of this agent is in a space determined by the dimensions of recent volatility and volume of the market, since anecdotal evidence suggest these factors are important in determining the "mood" of a stock market. In this case the agent has developed three distinct clusters, roughly corresponding to high volatility, high volume and low volume (both at medium levels of volatility). The models are most quite simple, e.g. predicting the price by looking at a price last week.

9.6 Conclusion

Culture is structured in a fundamental manner by social context. Any attempt to understand social integration or promote it will need to take this into account and not just as an afterthought. This chapter aims to convince that this is the case and indicate some of the ways forward in terms of simulation modelling, understanding and promotion. Such advances in modelling need to be part of a broader range of approaches alongside studies to gather evidence as to how and when social contexts "collide". Without good evidence to validate simulation models at many different levels (Axtell and Epstein 1994) one is limited to exploring abstract possibilities only.

Acknowledgements The research was done under grants GR/T11760/01 and EP/H02171X/1, both from the EPSRC. Its support and that of the MMUBS is gratefully acknowledged.

References

Alam, S. J., Geller, A., Meyer, R. and Werth, B. (2010). Modelling Contextualized Reasoning in Complex Societies with "Endorsements". *Journal of Artificial Societies and Social Simulation* 13(4)6 <http://jasss.soc.surrey.ac.uk/13/4/6.html>

Andrighetto, G., Campennì, M., Conte, R., and Cecconi, F. (2008). Conformity in Multiple Contexts: Imitation Vs Norm Recognition,. In *World Congress on Social Simulation 2008* (WCSS-08) George Mason University, Fairfax, USA.

Axtell, R. L. and J. M. Epstein (1996). *Growing Artificial Societies – Social Science From the Bottom Up*. Brookings Institution Press and MIT Press.

Axtell, R. L. and J. M. Epstein (1994). Agent-based Modelling: Understanding our Creations. *The Bulletin of the Santa Fe Institute***9**: 28–32.

Barwise, J. and Perry, J. (1983). *Situations and Attitudes*. Cambridge: MIT Press.

Conte, R. et al. (2010) EMIL-T: The final report of the EU 6FP EMIL Project. http://cfpm.org/cpmrep206.html

Cohen. P. R. and Grinberg, M. R. (1983) A Theory of Heuristic Reasoning About Uncertainty, AI Magazine, 4(2) http://www.aaai.org/ojs/index.php/aimagazine/article/view/393/329

Deacon, T.W. (1998) *Symbolic Species: The Co-Evolution of Language and the Brain*. Norton & Co.

Edmonds, B. (1998). Modelling Socially Intelligent Agents. *Applied Artificial Intelligence*, 12, 677–699.

Edmonds, B. (1999a) The Pragmatic Roots of Context. CONTEXT'99, Trento, Italy, September 1999. *Lecture Notes in Artificial Intelligence*, 1688:119–132.

Edmonds, B. (1999b). Capturing Social Embeddedness: a Constructivist Approach. Adaptive Behavior, 7:323–348.

Edmonds, B. (2001) Learning Appropriate Contexts. In: Akman, V. et. al (eds.) Modelling and Using Context – CONTEXT 2001, Dundee, July, 2001. *Lecture Notes in Artificial Intelligence*, 2116:143–155.

Edmonds, B. (2009) The Nature of Noise. In Squazzoni, F. (Ed.) Epistemological Aspects of Computer Simulation in the Social Sciences. *Lecture Notes in Artificial Intelligence*, 5466:169–182.

Edmonds, B. (2010a) Bootstrapping Knowledge About Social Phenomena Using Simulation Models. *Journal of Artificial Societies and Social Simulation* 13(1)8. (http://jasss.soc.surrey.ac.uk/13/1/8.html)

Edmonds, B. (2010b) Agent-Based Social Simulation and its necessity for understanding socially embedded phenomena. CPM Report No.: 10–205, MMU. (http://cfpm.org/cpmrep205.html)

Edmonds, B. (2012) Complexity and Context-dependency. *Foundations of Science*. Online First. DOI 10.1007/s10699-012-9303-x

Edmonds, B. and Moss, S. (2001) The Importance of Representing Cognitive Processes in Multi-Agent Models, Invited paper at Artificial Neural Networks – ICANN'2001, Aug 21–25 2001, Vienna, Austria. Published in: Dorffner, G., Bischof, H. and Hornik, K. (eds.), *Lecture Notes in Computer Science*, 2130:759–766.

Edmonds, B. and Moss, S. (2005) From KISS to KIDS – an 'anti-simplistic' modelling approach. In P. Davidsson et al. (Eds.): Multi Agent Based Simulation 2004. Springer, *Lecture Notes in Artificial Intelligence*, 3415:130–144.

Edmonds, B. & Norling, E. (2007) Integrating Learning and Inference in Multi-Agent Systems Using Cognitive Context. In Antunes, L. and Takadama, K. (Eds.) *Multi-Agent-Based Simulation VII, Lecture Notes in Artificial Intelligence* 4442:142–155.

Gärdenfors, P., The pragmatic role of modality in natural language. in 20th Wittgenstein Symposium, (Kirchberg am Weshel, Lower Austria, 1997), Wittgenstein Society.

Gilbert, N. (2006) When Does Social Simulation Need Cognitive Models? In Sun, R. (ed.) *Cognition and Multi-Agent Interaction: From Cognitive Modeling to Social Simulation*. Cambridge University Press, 428–432.

Granovetter, M. (1985) Economic Action and Social Structure: the Problem of Embeddedness., *American Journal of Sociology*, 91:481–93

Greiner, R., Darken, C. & Santoso, N.I. (2001) Efficient reasoning. *ACM Comp. Surveys*, **33**(1):1–30.

Hayes, P. (1995). Contexts in Context. Context in Knowledge Representation and Natural Language, AAAI Fall Symposium, November 1997, MIT, Cambridge.

Kokinov, B. and Grinberg, M. (2001) Simulating Context Effects in Problem Solving with AMBR. in Akman, V., Bouquet, P., Thomason, R. and Young, R.A. eds. *Modelling and Using Context*, Springer-Verlag, 2116:221–234.

Kummer, H., Daston, L., Gigerenzer, G., & Silk, J. (1997). The social intelligence hypothesis. In P. Weingart, S. D. Mitchell, P. J. Richerson, & S. Maasen (Eds.), *Human by nature. Between biology and the social sciences* (pp. 157–179). Mahwah, NJ: Erlbaum.

Schlosser, A., Voss, M. and Brückner, L. (2005). On the Simulation of Global Reputation Systems. *Journal of Artificial Societies and Social Simulation* 9(1)4 <http://jasss.soc.surrey.ac.uk/9/1/4.html>.

Simon, H. (1976). *Administrative Behavior* (3rd ed.). New York: The Free Press

Tykhonov, D, Jonker, C, Meijer, S and Verwaart, T (2008), Agent-Based Simulation of the Trust and Tracing Game for Supply Chains and Networks. *Journal of Artificial Societies and Social Simulation* 11(3) 1 http://jasss.soc.surrey.ac.uk/11/3/1.html.

Ye, M. & K. M. Carley (1995), Radar-Soar: Towards An Artificial Organization Composed of Intelligent Agents, *Journal of Mathematical Sociology* 20(2–3): 219–246.

Chapter 10
The Role of Stability in Cultural Evolution: Innovation and Conformity in Implicit Knowledge Discovery

Joanna J. Bryson

10.1 Introduction

Innovation is a topic of great interest in the study of cultural evolution (O'Brien and Shennan 2009). How do new behaviours and ideas come to be established in a culture? The reason for this interest is obvious—culture is after all an amalgamation of past innovations, so the study of innovation is also the study of the origins of culture. However, the emphasis on novelty that the term 'innovation' elicits may not be the most useful perspective for truly understanding culture origins. For evolution, the greatest challenge is *preserving* useful traits. The most essential characteristic of life is its capacity to reproduce—diversity and even increasing complexity, while also fascinating, occur in other materials as well.

How difficult is preserving culture? One indication of this may be the fact that so few species do to any measurable extent (Tomasello 1999). While we know that many species use culture as a part of their adaptive repertoire, there is little evidence for accumulation across generations (Whiten and van Schaik 2007). For some this has been seen as clear evidence that human culture is not so much a result of unsupervised processes akin to Darwinian evolution, but rather that it requires explicit and special mechanisms for transmission (Castro et al. 2004). Of course, this argument is either creationist or cyclical. While the current importance of culture to human survival means that any innovation for preserving innovation is highly adaptive biologically as well as culturally, it is unlikely that the initial innovation that supported cultural accumulation was acquired culturally (Bryson 2009, 2010).

Sperber and Hirschfeld (2004, 2006) argue that due to the noise inherent in the social transmission of behaviour, only a modular model of learning and mind can

J.J. Bryson (✉)
University of Bath, Bath, BA2 7AY, UK
e-mail: j.j.bryson@bath.ac.uk

V. Dignum and F. Dignum (eds.), *Perspectives on Culture and Agent-based Simulations*, Studies in the Philosophy of Sociality 3, DOI 10.1007/978-3-319-01952-9_10,

explain cultural preservation and stability. They propose that the massive modularity hypothesis (MMH) (Samuels 1998; Carruthers 2005) is an alternative to the current emphasis by many evolutionary anthropologists and behavioural ecologists on imitation as the source of culture (Huber et al. 2009). Unlike the modularity of Fodor (1983), under the MMH modules are acquired during development from culture. For example, the simple fact that a unique word exists is a clue for the related existence of a useful concept, and a child will search for a robust application of a novel term (Waxman and Markow 1995). This can apply similarly to more complex cultural input such as stereotypes (Clark and Kashima 2007) or myths (Steadman and Palmer 1997).

In this chapter I demonstrate through simulation the robustness of imitation learning given the context of a modular culture. This is simply because errors in transmission tend to cancel each other out provided there are sufficient exemplars and there is no bias in their production. I then go on to show however that a bias towards conformity is absolutely essential for preserving innovation, and in fact demonstrate the difficulty of constructing a model that can discover optimal new behaviours. At least for the simple abstract simulations presented here, it is difficult to find a single set of parameter values that both allow for innovation and preserve good solutions once found. I suggest that this may explain why cultural species do not tend to have a single such set of parameters, but rather to be neophilic and neophobic at different ages or in different contexts.

10.1.1 Terms and Concepts: Cultural Evolution and Innovation

Whether culture can be usefully thought of as an evolutionary system is a matter of ongoing debate (Aunger 2000; Richerson and Boyd 2005; Wimsatt 2009). While few doubt that the biological capacity for culture must have evolved, the question is whether or not culture *itself* evolves. That is, are the contents of cultures themselves (e.g. words, ideas, symbols, images or even just socially-acquired behaviour) subject to reproduction, variation and selection in a way that is meaningfully similar to the process Darwin identified as explaining the origin of species. While acknowledging this controversy, in the present chapter I will not address it directly, but rather just assume an evolutionary perspective towards culture. This is a standard approach for simulation: to the extent that any results are validated by comparison to their target system in the natural world, these results can be seen as also verify the initial axiomatic assumptions behind the simulation as well (Bryson et al. 2007).

Taking then a selectionist perspective, we might usefully view innovation as errors in the cultural replication and preservation process that happen to persist. Of course this perspective is a simplification. There may well be intelligent search performed by some individual ‘carrier’ of the culture that is the root cause of some specific ‘defect in replication’. Cultural evolution is not *necessarily* an unsupervised and unintentional search process. Further, there is no reason for inheritance in cultural evolution to be limited to one or two parents and a single recombination event (Bryson 2008, pp. 89–90). Rather, the more information that can be gathered, the easier it is to detect the salient signal inside the noise and irrelevant detail. Finally, any particular variation in culture may actually convey a *biologically*-adaptive benefit, so it may not just 'happen' to persist. However, taking a simplified meme's-eye view of innovation may help us understand the processes that underly cultural change (Dawkins 1976).

I take it as given that *some* cultural variation happens as a result of blind chance and copying errors. For the sake of simplicity therefore, this will be the only sort of 'invention' in the models described here. Presumably intelligent invention only accelerates the pace of change by making actually adaptive 'errors' more frequent, but otherwise does not substantially alter the process. In an effort to keep this chapter as clear as possible, I will call any deviation from a previously-established culture an invention, and any invention that reliably persists through cultural transmission an innovation. The models below show conditions where an adaptive innovation can be made, and conditions where innovations occur even though they have no adaptive impact.

10.2 Background: Modularity and Cultural Stability

It is useful to decompose the social communication of behaviour into two different levels. For the purpose of this chapter, I will refer to the rote replication of end effector positions or end effects as *imitation*. This is a simplistic, ordinary-language use of the term, but sufficient for the experiments described here.[1] By 'imitation' I do *not* necessarily mean a full transfer of behaviour. This latter would imply that two agents have communicated not only actions but a model between them, such that they have the same understanding of the role of the actions they imitate, and the goals they might meet with those actions. Sperber and Hirschfeld argue that this shallow sort of imitation cannot be an integral part of cultural transmission. Although I generally find their work very useful, my main departure is that I believe shallow imitation *can* fulfill this role.

Sperber and Hirschfeld (2004, 2006) argue that due to the unreliability of both performing actions and perceiving others' acts, reliable cultural transmission is exceedingly unlikely. Giving evidence based on the known degradation of signal experienced in simple transmission chains of spoken sentences (e.g. the party game of Telegraph *[USA]* or Chinese Whispers *[UK]*), they criticise the current emphasis on the role of imitation in cultural transmission. Imitation is limited to mere replication of apparent behaviour, and that is in turn limited by constraints in our ability to perceive others' actions, and also by our own inability to execute our actions exactly was we intend. Sperber and Hirschfeld insist that what matters is the

[1]For more elaborate definitions see e.g. Whiten and Ham (1992) and Bryson (2009).

deep transfer of mental models from one mind to another, not the shallow imitation of expressed behaviour.

How can this deep model be recovered from limited perceptual information? Sperber and Hirschfeld see no way, and use this implausibility as evidence that some information must come from elsewhere. They suggest this missing information is the information encapsulated in modules. Modules under the MMH may have both genetic and explicitly-learned components. Thus extra information is available to compliment the shallow information available from perception and imitation.

People used to implementing artificial learning systems and/or familiar with the mathematics or logic of learning may find the above arguments somewhat unsatisfying. After all, provided that errors in perception or action are random, they can be considered noise and will cancel each other out provided there are enough observations. Anything left is not random and is therefore also not noise, but rather some sort of signal which ought to be useful. However, this only accounts for part of the Sperber and Hirschfeld argument, and the other part (that some information is missing) I think is correct, though their theory is under specified. Where does the extra information they postulated as coming from modules *itself* originally come from? Biological evolution, cultural evolution and individual learning are all forms of learning. Therefore taken as sources of information and knowledge, their power is essentially identical (Wolpert 1996b; Best 1999). Thus to some extent the Sperber and Hirschfeld argument is overly compartmentalised. To say that the extra information required to make sense of the noisy social transmissions *comes from* modules is still to beg a question of how the modules themselves have come to support this process.

Although they are not completely explicit about it—in fact, they are almost explicitly agnostic on the topic (Sperber and Hirschfeld 2004, p. 41)—it seems likely Sperber and Hirschfeld are implying that some of what we commonly call 'human culture' is genetically encoded. This is problematic if we take the simple information-centred definition of culture I ordinarily favour: that culture is all behaviour acquired from conspecifics by non-genetic means (Richerson and Boyd 2005; Bryson 2009). However, taking instead a more ordinary-language view of culture as the aspects of behaviour such as language and social organisation which seem to vary between peoples, then the idea of a genetic component becomes more sensible. There is relatively little controversy for example that *some* aspects of linguistic competence must be genetic (such as the capacity for vocal imitation and transmission), though others are clearly learned by individuals from their own or another culture (Fitch 2005). Given what we understand about how learning affects evolution (Baldwin 1896; Hinton and Nowlan 1987; Borenstein and Krakauer 2008), we should also expect that some things that may first evolve as cultural variation could over time become at least partially genetically entrenched.

10.2.1 Modularity and Learning

What Sperber and Hirschfeld really propose then is that the automatic or implicit learning of culture from imitation cannot in itself account for all the richness of human culture. Although they acknowledge a possible complementary role for imitation-driven cultural transmission, their own emphasis is on complex mental models underpinning human behaviour. This process in turn requires the explicit transfer of abstract/symbolic knowledge. Symbols in themselves contain almost no information, but cultural participants who understand them have high-information-content associations, or *grounding*, for them. Under the Sperber and Hirschfeld model, grounding is encoded in modules and contains most of the information necessary for the newly acquired behaviour.

This notion of the role of modules is quite similar to one I have proposed in the context of artificial intelligence (Bryson 2000, 2001). In that work I extended the model of modular organisation of intelligence known as Behavior Based Artificial Intelligence (BBAI) (Brooks 1991) to include module-based learning. The original insight of BBAI was that real-time intelligence is best decomposed into behaviour modules. 'Best' in this context means

- Responsive to the demands of an unpredictable and rapidly changing environment,
- Robust to the difficulties of both sensing and control, and
- Easily and reliably developed by programmers and roboticists.

Under standard BBAI, the purpose of a behaviour module is to perform some action or provide some capacity for its host agent. Modules consist therefore of instructions for whatever control is necessary for those actions, but also of whatever perception is necessary to guide those actions. This tight coupling of sensing to action is a hallmark of BBAI. It simplifies the problem of building intelligence by restricting the problems worked on to a minimum set of capacities each with only the most essential detail required to reliably execute its tasks. The strength of the approach was not only argued but also demonstrated in the first robots able to move autonomously at animal-like speeds (Horswill 1993; Brooks 1990).

The Bryson (2001) extension to BBAI stems from the observation that perception is more than just sensing. At any one instant, sensing provides too little information to successfully disambiguate the correct next action. Animals address this problem through systems of memory ranging from integrating recent signals through conventional ideas of memory (e.g. map learning) and on through genetically provided biases (Carlson 2000; Rao 1999). This applies to BBAI as well. Just as behaviour modules should contain the dedicated and specialised sensing necessary for their actions, they should also contain the dedicated and specialised memory necessary for both perception and control. One advantage of this modularisation of learning is that specialised representations can be chosen that facilitate the particular sort of learning that each module needs. This increases the probability that the individual agent will learn and act successfully (Wolpert 1996b).

10.2.2 Bootstrapping Culture: The Law of Large Numbers

From the above review it should be obvious that I strongly support the idea that modules can and almost must support all learning. Strictly speaking, modular learning systems can always be recast as a computationally-equivalent homogenous ones. That is in theory a homogeneous system can learn anything a modular one can (Wolpert 1996a). However, accurate learning is much, much less probable without the bias which modularity can provide, and therefore will take much longer on average to converge (Bishop 2006). For an animal or other real-time system, this means learning is less likely to succeed in time to be used.

This result includes the individual learning that underlies cultural transmission and evolution. However, we must consider the full process of internalising information to guide behaviour, from evolution through development and learning. We also need to account for cultural transmission in the non-human species in which it has been observed (Whiten et al. 1999; van Schaik et al. 2003; Perry and Manson 2003; Kenward et al. 2006; Dornhaus and Franks 2008; Wilkinson et al. 2010). Even ants might be thought of as having minor cultural differences between colonies, since their members both determine and learn new nest locations in a distributed, social manner (Franks and Richardson 2006).

Sperber and Hirschfeld are correct to be skeptical of one-shot imitation as a mechanism of social transmission. Essentially, if a single signal can transmit enough knowledge to really alter behaviour, then that knowledge must have been previously accumulated and stored in such a way that the behaviour observed has information-equivalence to a symbol anyway (Wood 2008). In this case, imitation is not fundamentally different from explicit communication. There will in fact be a continuum of conditions whereby true communication of cultural contents can be achieved with more or less information prompted, depending on how much the cultural and genetic predispositions of the demonstrator and the receiver align. To return to the telegraph metaphor, the way real telegraphs work is through a system of repeaters that can remove noise accumulated and re-boost the signal. Where the repeating process is intelligent, degradation is probably even less of a problem.

10.3 Experiment 1: Stability of Culture with Noisy Transmission

The following experiments demonstrate the above arguments, and then move to explore some of their consequences. They are abstract and I have not been able to think of a good way to validate them, so at this stage of development they should probably be thought of as no more than intuition pumps (Dennett 1995). Here I present a modular model of a culture. The model is agent-based (ABM). It is built in NetLogo, a standard and freely-available ABM development environment

(Wilensky 2011). The code for the model is available from the author by request, or her Web site by demand.

10.3.1 Model

An ABM consists of three parts (Bryson et al. 2007):

1. An *environment* where the agents are situated and which determines their possible behaviour;
2. *Attributes*, also known as parameters or variable state, which describe the agents and what makes them individual; and
3. *Behaviour* or intelligence, the actual algorithms which the agents use for control.

I describe each of these in turn.

Environment The first model has a very simple environment. It is entirely social, with no intrinsic reward provided for any behaviour. Space is described as a torus—that is, a square with the left and right edges connected, and also the top and bottom ones. This means that the code and analysis do not have to deal with exceptional agents that live at the edge of their world. Agents occupy every possible location in the grid; each has eight neighbours it can observe.

Agent Attributes Agents have three types of attributes (Bryson et al. 2007):

1. *Static parameters* which vary only between experimental conditions,
2. *Run-dependent parameters* which vary per run and often per individual but are fixed at the beginning of the run, and
3. *Dynamic parameters* which change within a single agent's lifetime.

Besides having eight neighbours, the most fundamental static parameter in this model is the agents' modules. All agents have the same number of modules. Although the exact number of modules is run-dependent, how they operate is static. Each module is very simple—it is intended to correspond to a context the agent may find itself in. Each agent has a single behaviour that it currently expresses in that context; *which* behaviour among many possible is learned socially (see algorithm below). For convenience in visualisation (but not in explication) there are exactly as many possible behaviours for each context/module as there are modules.

Since the agents acquire their behaviour socially, they need to be able to keep track of other agents' behaviour they witness. Thus each agent has associated with each module a memory. The size of this memory is the same as the number of possible actions. The agent remembers how many times it has seen each action it has witnessed in each context. Thus the content of this memory is a dynamic parameter.

Besides the contents of its memory, the only other dynamic parameter of an agent is its age. At the very beginning of a simulation, age is assigned randomly to each agent from the full range of possible values. Subsequently, any new agent starts with age 0.

In addition to the number of modules, there are a number of other run-dependent parameters:

- Each agent's (X, Y) position in social space. This determines which other eight agents are its neighbours.
- The number of 'years' spent as a child and as an adult. The difference is that no one learns socially from children.
- The number of acts performed per 'year'. This in combination with the lifespan and the size of the culture determines how much each agent will experience in its 'life'.
- The probability of a perception error and the probability of an action error. If one agent performs an action error, all of its neighbours will see an unintended behaviour in a particular context. If one agent experiences a perception error, then it is the only agent that's knowledge is affected. In both cases, an error means a value for an action is randomly drawn from all possible acts. For the sake of simplicity, in the experiments discussed here the only probability varied was of action error. This is more likely than perception error to cause perturbations of culture, since it can bias eight neighbouring agents' beliefs in the same way.

  This variable is somewhat dynamic, in that it can be varied during the course of a simulation by the experimenter. This allows for the experimenter to search for a threshold value below which the culture is stable, and above which the culture degrades. However, nothing the agents do themselves changes this value, so from their perspective it is run-dependent.
- The weight given to the seed culture at the beginning of the simulation. At the beginning of the simulation, all of the first generation of agents have their memories set to an initial cultural value. For each context in the initial culture, the favoured behaviour is the one with the same index as the context. Thus for context 0 the favoured behaviour is 0, for context 1 it is 1, and so forth. The value weighting the favoured behaviour is set by the experimenter. If the weight is five, the agents have a memory equivalent to having seen other agents perform that action five times. This parameter has no other role in the simulation after the first generation has died.

For visualisation, the field of agents is visible as a square. The agents are arrow shaped. The agents are coloured to indicate their age: children are light and adults dark. The viewer can be set to examine any one behaviour context for all the agents. The beliefs and therefore the chosen action of each agent for that context is then visualised as the angle at which the agent points. The angle $= (360 * i)/N$, where i is the number of this particular belief, and N is the number of possible beliefs. As a secondary visualisation, there is also a chart which shows the percentage of agents that conform to their original beliefs in the seed culture for the first four contexts. Since all contexts are functionally identical, these first four can be treated as a small random sample of all the modules.

Agent Behaviour On every program cycle, a context is chosen by the environment at random. Each agent then checks its memory for that context and expresses whatever action it has itself most often witnessed in that context. If more than one action is tied for having been witnessed the greatest number of times, then the tied actions are chosen between at random. Assuming there is some Probability of Action Error (PAE), the agent then has a PAE chance of choosing an action randomly from all possible values and expressing it. Otherwise, it expresses its module's true value.

"Expressing an action" in the simulation is manifest as an agent asking all eight of its neighbours to add one count to that action's value in that context, indicating that action/context pairing has been witnessed once more. If there were a probability of perception error, at this point a random value might be introduced into an individual's memory rather than the act expressed. However it is best practice to limit the number of parameters on a model for simplifying analysis, and for the reason stated above I chose only to manipulate action errors for the experiments presented here.

When an agent reaches its age limit, it dies. When an agent dies, it is immediately replaced with a new agent of 0 age. This new agent has a completely empty mind. It has the same number of modules as the rest of the agents in the simulation, but every possible value for every module is given 0 weight. Thus its initial actions will be entirely random.

10.3.2 Results

Cultural stability is directly correlated to the number of exposures to an action that an agent is likely to experience for each action in its lifetime. Thus the longer adult life, and the more actions that occur per year, the more stable culture. On the other hand, having more modules decreases the number of actions *per* module, so this is negatively correlated to stability, as of course is the PAE.

One surprising result concerns the influence of children on culture. The tendency to ignore children's behaviour (which is initially essentially arbitrary) has been proposed as a mechanism of cultural stability. However, because even children after 1 year are more likely to express their culture's values for any module than any other value, shortening "childhood"—or at least, the period where children do not serve as cultural models—actually *increases* cultural stability. Of course disrespect is not the only attribute of childhood. If I had modelled childhood also as a period when more is time devoted to observation of others (perhaps by increasing the neighbourhood size for children), then a longer childhood might have been more beneficial.

Figure 10.1 shows a run with parameters set such that the culture is fairly stable, but not sufficiently so to stop degradation (forgetting) of the culture. Since we are observing the $i = 0$ context module, the agents conforming to the original culture are pointing straight up. Notice that young agents (the light/yellow agents) may be oriented in any direction since they will not have seen many expressions

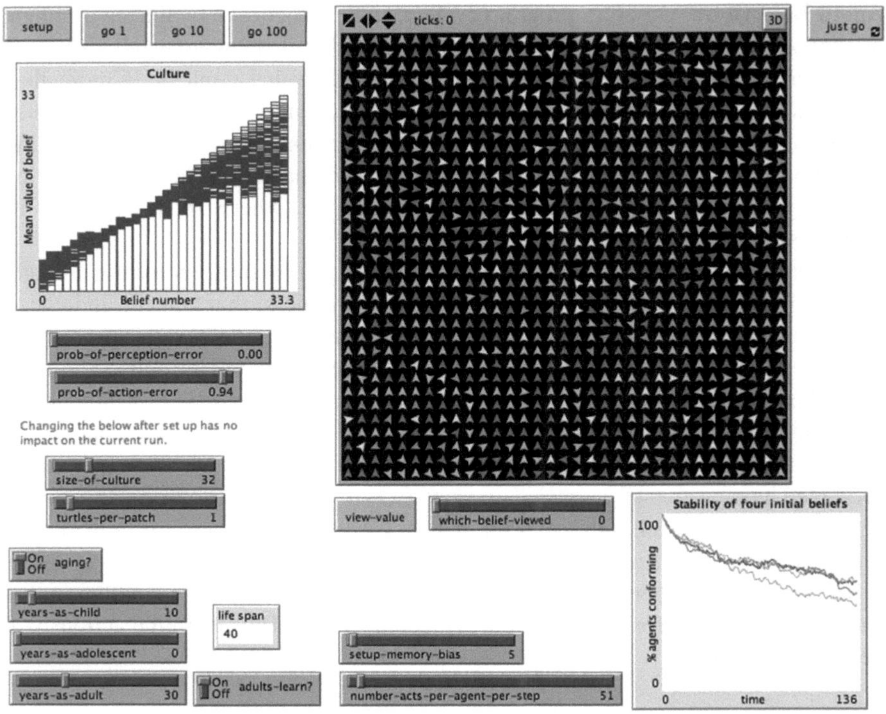

Fig. 10.1 Culture degrading. Notice the presence of subcultures among neighbouring adults

of behaviour in this context yet. However, where adults (dark/blue agents) are misoriented, they often are so in company. Thus the same mechanisms that largely preserve culture can also serve to form and preserve subcultures.

Figure 10.2 shows the same simulation in the future. However, just after the previous snapshot, the probability of action error was lowered from 94 to 90 %. Notice this does not simply freeze the decline of the culture, but actually results in the initiation of a rapid recovery. This is because the level of conformity to the original culture was still $>1/N$. If culture had degraded to total chaos, then reducing the PAE would have lead to conformity as well, but not necessarily to the original value. Note also that a culture will never have 100 % conformity because of the ignorance of children, but with a low PAE a stable culture will achieve a high level of conformance.

10.3.3 Discussion

The idea that a module might take only a few discrete values may seem such an extreme abstraction that it renders the model meaningless. However, we know that

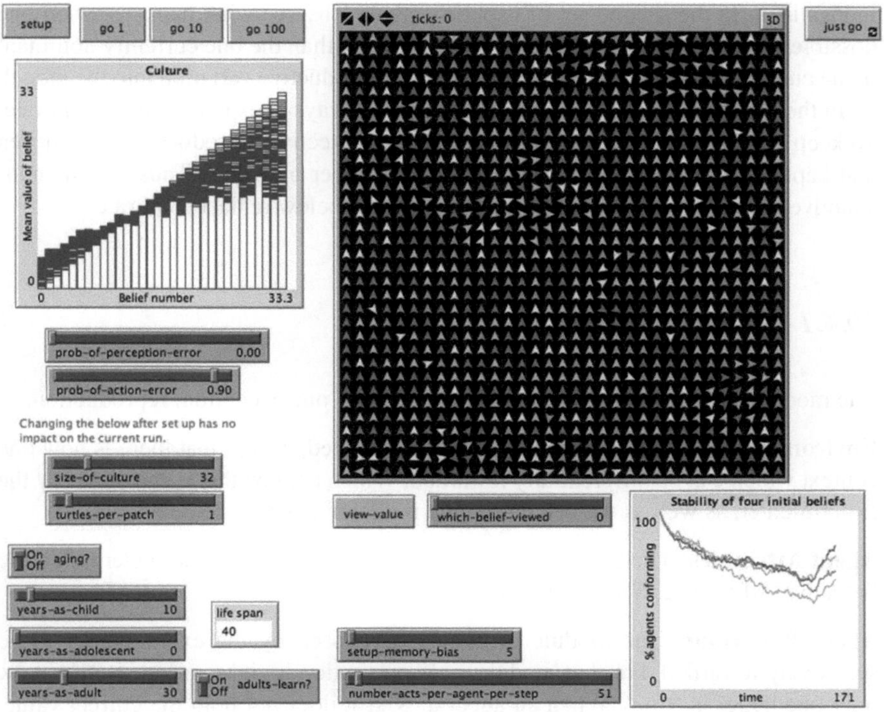

Fig. 10.2 Culture recovering. The probability of generating incorrect actions has been reduced just 4 %

animals including humans are extremely inclined to categorise perceptual data. Even in continuous domains such as the light spectrum, humans are far more sensitive to variation near the "boundaries" between named colours than well within them (Harnad 1987; Steels and Belpaeme 2005). This emphasises the role both Sperber and Hirschfeld and I hypothesise for modules in learning in general, of which social learning is a special case. Through some combination of genetics and experience the agent is assumed to know a set of categories or concepts, which learning facilitates a choice between.

Social learning may also facilitate the discovery of *new* categories and modules by signalling through variations in behaviour a perceptual difference an agent had not otherwise detected (Bates 1999; Bryson 2008). However, module construction is not modelled in the current simulations.

10.4 Experiment 2: Innovation

In the first model we already witnessed the formation of subcultures. Since these can be stable for a few years or even generations, they might already be viewed as innovations. In the second set of experiments we observe what happens when one possible value for a culture model is more adaptive than the one currently dominant in the culture. To do this, we have to introduce reproductive variation into the model.

In the previous simulation, reproduction was always at exactly replacement rate. To keep the experiment simple, a mechanism of selective reproduction was chosen that kept a full environment as the *maximum* number of agents. Thus, for the non-adaptive culture values, reproduction was lowered below replacement rate.

10.4.1 Model

The model is largely as described before, with only one exception: reproduction.

Environment The environment is largely unchanged, except that there is now one context which can be differentially rewarded. Which context this is can be set by the experimenter, as well as the behaviour that is adaptive in that context.

Agent Attributes There is one new attribute, a run-dependent parameter reflecting Selective Advantage, SA, described below.

Agent Behaviour One module or context is chosen by the experimenter to be selectively rewarded. For that module, only one action is right, or put another way, only one belief is 'true'. When an agent dies, if it does not hold the correct value, then its probability of being replaced is reduced by SA. On the other hand, if an agent does have the adaptive belief, not only will it certainly reproduce, but also if one of its neighbouring spaces is available, it will create one additional offspring. This allows the recovery of the population.

Note that because all agents are identical, there is no change in *genetic* distributions due to this advantage. What a parent leaves to its child or children is only its neighbours—its social network.

10.4.2 Results and Discussion

Surprisingly (and ironically), my explorations of the parameter space have shown that a culture needs to be strongly disposed towards stability in order for a new tradition to take root. If culture degrades easily, then even when agents stumble on the adaptive subculture they forget it again within a few generations. Obviously, however, it takes considerable disruption for a stable culture to lose its existing values so it can change to the adaptive ones. As the model is currently built, this disruption takes the form of the loss of neighbours and therefore the lower probability of discriminating the cultural values accurately. When one isolated subculture does stumble on the adaptive value and begin refilling the space around it, then the propensity for stability returns.

If the culture parameters are set to a lower level of stability, then the dominant culture can stop dominating earlier, but any new subculture has significantly more difficulty becoming self-maintaining. The adaptive subculture in particular becomes surrounded by juveniles (that is, relatively young agents) filling empty spaces. Due to less prior experience, relatively young agents are more open to influence—both to random patterns of other juveniles and to the influence of members of other neighbouring subcultures. Because it will still be disproportionately wide-spread in the culture, the ring of juveniles is particularly vulnerable to invasion by the original, non-adaptive value held by that culture. Since they surround the core of 'true' (adaptive) believers, they will generally sway their behaviour and the true belief is lost.

Another significant factor determining the outcomes for this simulation is the probability of stumbling on the correct answer in the first place. Recall that in all these simulations all behaviours are equally probable for naïve agents. If there are too many possible values for the module that is subject to selection, the agents are unlikely to find the rewarded value in time to save themselves from extinction. If the simulation were changed so that the agents were even slightly more intelligent in their search—for example, if they could remember neighbours that failed to reproduce or succeeded in having two children, this would increase the probability of the correct action being chosen.

Another interesting result is that although only one module was subject to selective pressure, the cultural norms for other modules also change. This might be because the same agents that are likely to discover the adaptive innovation had a general tendency for invention. Although all the agents have identical programs and are seeded randomly at the beginning of the simulation, the population is not entirely homogeneous. Chance patterns of distribution of age—the only differentiation between agents in the initial population—can lead to some patches of space being more or less likely to deviate from the cultural norm and form a subculture. Due to the policy of reproduction by replacement, age patterns are fairly stable. Another explanation is that change simply occurs due to the drop in cultural stability with the reduction of numbers. However, since the other modules are not having their original culture actively selected against, in some cases they recover their original value after the population stabilises (see chart in lower right of Fig. 10.4).

Another unanticipated result from this experiment was that the pattern of regrowth after the adaptive behaviour was discovered lead to large regions of adjacent age cohorts. This in turn seems to lead to the emergence in many but not all of the module contexts not subject to selection of multiple stable cultures. Figure 10.4 shows an example of one such. This may have analogues in natural culture, where age cohorts may communicate predominantly internally rather than mixing with other ages. Even where there is a mix of ages, it is possible for age cohorts to focus their social learning attention on their peers.

The figures show a run where the PAE was set to what was in the non-selective condition a fairly stable value, particularly given the number of modules in the culture. Figure 10.3 shows the cultural values for the context and module subject to selective pressure when the number of agents holding the adaptive belief has just

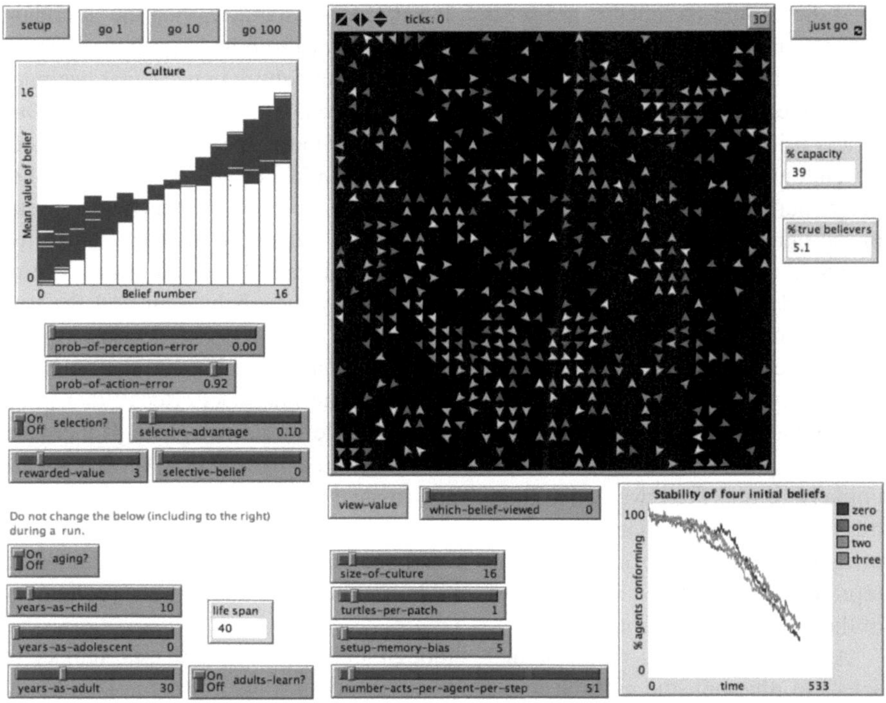

Fig. 10.3 The point in time when an adaptive innovation is just beginning to dominate a culture

begun to outnumber those conforming to the original culture. Figure 10.4 shows the same run after the population has recovered. This second figure observes not the context subject to selection, but one of the other contexts where the values are arbitrary from a selective perspective. This context has now formed multiple sizeable, stable subcultures. Notice the pattern of ages in the agents as indicated by their colour.

10.5 Conclusions and Discussion

In this chapter I have examined and to a large extent supported the proposal of Sperber and Hirschfeld (2006), while at the same time clarifying some details of how their system might work. The modules they describe utilise information previously acquired either by the species (encoded genetically) or by the individual's learning, which of course may also be canalised by the species through culture.

The model I have presented demonstrates the ability of a culture to be stable in the face of enormous errors in communication. The famous 'poverty of the stimulus' is simulated by the high level of noise in the actions actually generated by the agents. Agents are nevertheless able to derive a signal because of the Law of Large Numbers

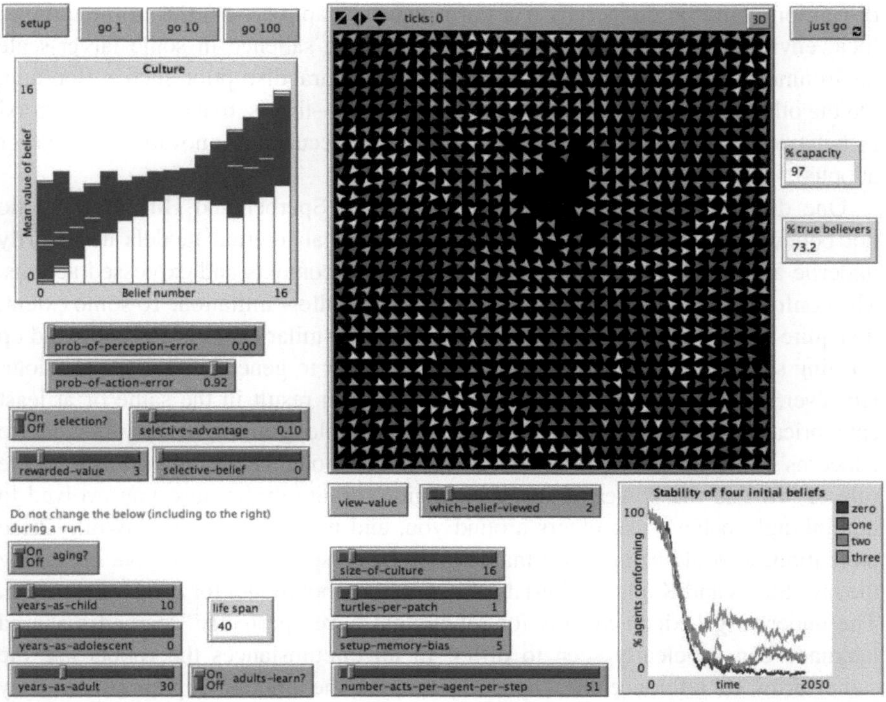

Fig. 10.4 The impact of adaptive selection for a new value in one module on the cultural values of another. The module governing the behaviour shown has no ‘true’ (adaptive) belief; sub cultures for this behaviour have emerged as a result of the social dynamics resulting from selection on another behaviour

and the fact the errors are unbiased. In these simulations all behaviour contexts are equally probable and all social demonstrations equally salient. In human culture we know that rare but important cultural behaviours such as rituals tend to be associated with high emotion salience indicators such as music which may assist in emphasising particular memories (LeDoux 1996; Whitehouse 2002). For example, in medieval England the relatively boring and seldom-performed but essential task of patrolling the parish boundaries was made salient to young boys by beating them at boundary stones so the boys would remember the stones’ locations (Darian-Smith 2002).

The models also show circumstances in which innovations can not only take place but take hold. Strong tendencies towards conformity can give rise to small stable subcultures even in strictly arbitrary environments, as shown in Experiment 1. Experiment 2 explores the conditions necessary for acquisition of a newly-adaptive norm—that is, an action selected by the environment. In addition, it also shows that society-wide displacements of one cultural norm for another can take place for no direct adaptive reason, but simply as a side-effect of the disruption to the society necessary for another, more urgent change in cultural norms. This incidental disruption could be dangerous if a norm that is adaptively-neutral in the current, local environmental context actually held adaptive salience in some larger-scale environmental context, for example in times of a natural disruption such as flooding. On the other hand, if the society is too conservative—that is, makes too few 'errors' in behaviour replication, then inventions seldom occur and innovations are never adopted.

One difference between my work and that of Sperber and Hirschfeld—I do not believe they are correct to assume that identical internal models necessarily underlie apparently identical connections between contexts and expressed actions. The conformance demonstrated here is based on shallow imitation. To some extent, it is quite likely that agents with similar brains and similar experiences will wind up forming similar internal models or theories in order to generate similar behaviour. However, it is possible that multiple models would result in the same or at least categorically indiscriminable behaviour. For example, you might obey law due to concerns about an afterlife, due to an elaborate model of the importance of the rule of law and the power of social contagion, or simply because you evolved to unthinkingly behave like others around you, and most of them are lawful. These three models would be indiscriminable from the perspective only of your observing the law. Steels and Kaplan (1999) demonstrates a robot model for this phenomenon. The underlying lexicon models for robots that have "perfectly" learned a shared language can be clearly seen to differ. In all circumstances the robots use the same terms to reference the same objects, yet the internal representation they require for grounding the terms as mappings to their sensor and motor states vary considerably between robots. Thus model conformance is not a necessary part of social conformance, and may in fact provide a useful source of variation to the populations' inventions.

The simulations I have described beg much further analysis. For example there should be a more thorough exploration of the effects of developmental differences in communication on the adaptation of cultures to new circumstances or to the opportunities of adaptive innovations. Further, the spontaneous emergence of stable subcultures in both sets of experiments might be seen as examples of sympatric speciation—a process often attributed to sexual selection. Clearly no equivalent of sexual selection takes place here. Although the model is intended to be one of cultural evolution, it might easily be extended to model biological evolution to study this process. Or, we might hypothesise that cultural evolution underlies the beginning of sympatric speciation, and the process is then genetically consolidated. These projects are left as future work.

Acknowledgements This research was inspired by an informal talk by Dan Sperber in the Spring of 2008 at the Konrad Lorenz Institute for Evolution and Cognition Research, where I was supported by the institute as the Hans Przibram Fellow with sabbatical assistance from the University of Bath. Thanks to Christophe Heintz for his discussion and comments on the original version of this paper, which was also presented to *The Fall AAAI Symposium on Adaptive Agents in Cultural Contexts (AACC'08)*, and appeared in its informal proceedings (A. Davis and J. Ludwig, eds). Thanks to that symposium also for their comments. Effort on completing the final version was sponsored by the US Air Force Office of Scientific Research, Air Force Material Command, USAF, under grant number FA8655-10-1-3050.

References

Aunger, R., editor (2000). *Darwinizing culture: The status of memetics as a science*. Oxford University Press, Oxford.

Baldwin, J. M. (1896). A new factor in evolution. *The American Naturalist*, 30:441–451.

Bates, E. (1999). Plasticity, localization and language development. In Broman, S. and Fletcher, J. M., editors, *The changing nervous system: Neurobehavioral consequences of early brain disorders*, pages 214–253. Oxford University Press.

Best, M. L. (1999). How culture can guide evolution: An inquiry into gene/meme enhancement and opposition. *Adaptive Behavior*, 7(3/4):289–306.

Bishop, C. M. (2006). *Pattern Recognition and Machine Learning*. Springer, London.

Borenstein, E. and Krakauer, D. C. (2008). An end to endless forms: Epistasis, phenotype distribution bias, and non-uniform evolution. *PLoS Computational Biology*, 4(10):e1000202.

Brooks, R. A. (1990). Elephants don't play chess. In Maes, P., editor, *Designing Autonomous Agents: Theory and Practice from Biology to Engineering and Back*, pages 3–15. MIT Press, Cambridge, MA.

Brooks, R. A. (1991). Intelligence without representation. *Artificial Intelligence*, 47(1–3):139–159.

Bryson, J. J. (2000). Making modularity work: Combining memory systems and intelligent processes in a dialog agent. In Sloman, A., editor, *AISB'00 Symposium on Designing a Functioning Mind*, pages 21–30.

Bryson, J. J. (2001). *Intelligence by Design: Principles of Modularity and Coordination for Engineering Complex Adaptive Agents*. PhD thesis, MIT, Department of EECS, Cambridge, MA. AI Technical Report 2001–003.

Bryson, J. J. (2008). Embodiment versus memetics. *Mind & Society*, 7(1):77–94.

Bryson, J. J. (2009). Representations underlying social learning and cultural evolution. *Interaction Studies*, 10(1):77–100.

Bryson, J. J. (2010). Cultural ratcheting results primarily from semantic compression. In Smith, A. D. M., Schouwstra, M., de Boer, B., and Smith, K., editors, *The Evolution of Language 8*, pages 50–57, Utrecht.

Bryson, J. J., Ando, Y., and Lehmann, H. (2007). Agent-based models as scientific methodology: A case study analysing primate social behaviour. *Philosophical Transactions of the Royal Society, B — Biology*, 362(1485):1685–1698.

Carlson, N. R. (2000). *Physiology of Behavior*. Allyn and Bacon, Boston, seventh edition.

Carruthers, P. (2005). The case for massively modular models of mind. In Stainton, R., editor, *Contemporary Debates in Cognitive Science*, pages 205–225. Blackwell Publishing.

Castro, L., Medina, A., and Toro, M. A. (2004). Hominid cultural transmission and the evolution of language. *Biology and Philosophy*, 19(5):721–737.

Clark, A. E. and Kashima, Y. (2007). Stereotypes help people connect with others in the community: A situated functional analysis of the stereotype consistency bias in communication. *Journal of Personality and Social Psychology*, 93(6):1028–1039.

Darian-Smith, E. (2002). Beating the bounds: Law, identity and territory in the New Europe. In Greenhouse, C., Warren, K., and Merz, E., editors, *Ethnography in Unstable Places*, pages 249–275. Duke University Press, Raleigh, NC.

Dawkins, R. (1976). *The Selfish Gene*. Oxford University Press.

Dennett, D. C. (1995). Intuition pumps. In Brockman, J., editor, *The Third Culture*, pages 181–197. Simon & Schuster, New York.

Dornhaus, A. and Franks, N. R. (2008). Individual and collective cognition in ants and other insects (Hymenoptera: Formicidae). *Myrmecological News*, 11:215–226.

Fitch, W. T. (2005). The evolution of language: A comparative review. *Biology and Philosophy*, 20(2–3):193–203.

Fodor, J. A. (1983). *The Modularity of Mind*. Bradford Books. MIT Press, Cambridge, MA.

Franks, N. R. and Richardson, T. (2006). Teaching in tandem-running ants. *Nature*, 439(7073):153.

Harnad, S. (1987). Categorial perception: A critical overview. In Harnad, S., editor, *Categorial perception: The groundwork of perception*. Cambridge University Press.

Hinton, G. E. and Nowlan, S. J. (1987). How learning can guide evolution. *Complex Systems*, 1:495–502.

Horswill, I. D. (1993). *Specialization of Perceptual Processes*. PhD thesis, MIT, Department of EECS, Cambridge, MA.

Huber, L., Range, F., Voelkl, B., Szucsich, A., Virányi, Z., and Miklosi, A. (2009). The evolution of imitation: What do the capacities of non-human animals tell us about the mechanisms of imitation? *Philosophical Transactions of the Royal Society, B — Biology*, 364(1485):2299–2309.

Kenward, B., Rutz, C., Weir, A. A. S., and Kacelnik, A. (2006). Development of tool use in New Caledonian crows: Inherited action patterns and social influences. *Animal Behaviour*, 72(6):1329–1343.

LeDoux, J. (1996). *The Emotional Brain: The Mysterious Underpinnings of Emotional Life*. Simon and Schuster, New York.

O'Brien, M. J. and Shennan, S. J. (2009). *Innovation in cultural systems: Contributions from evolutionary anthropology*. MIT Press, Cambridge, MA.

Perry, S. and Manson, J. H. (2003). Traditions in monkeys. *Evolutionary Anthropology*, 12:71–81.

Rao, R. P. N. (1999). An optimal estimation approach to visual perception and learning. *Vision Research*, 39(11):1963–1989.

Richerson, P. J. and Boyd, R. (2005). *Not By Genes Alone: How Culture Transformed Human Evolution*. University Of Chicago Press.

Samuels, R. (1998). Evolutionary psychology and the massive modularity hypothesis. *The British Journal for the Philosophy of Science*, 49(4):575–602.

Sperber, D. and Hirschfeld, L. (2006). Culture and modularity. In Carruthers, P., Laurence, S., and Stich, S., editors, *The Innate Mind: Culture and Cognition*, volume 2, pages 149–164. Oxford University Press.

Sperber, D. and Hirschfeld, L. A. (2004). The cognitive foundations of cultural stability and diversity. *Trends in Cognitive Sciences*, 8(1):40–46.

Steadman, L. and Palmer, C. (1997). Myths as instructions from ancestors: The example of Oedipus. *Zygon*, 32(3):341–350.

Steels, L. and Belpaeme, T. (2005). Coordinating perceptually grounded categories through language: A case study for colour. *Behavioral and Brain Sciences*, 28(4):469–489.

Steels, L. and Kaplan, F. (1999). Bootstrapping grounded word semantics. In Briscoe, T., editor, *Linguistic Evolution Through Language Acquisition: Formal and Computational Models*. Cambridge University Press.

Tomasello, M. (1999). *The Cultural Origins of Human Cognition*. Harvard University Press, Cambridge, MA.

van Schaik, C. P., Ancrenaz, M., Borgen, G., Galdikas, B., Knott, C. D., Singleton, I., Suzuki, A., Utami, S. S., and Merrill, M. (2003). Orangutan cultures and the evolution of material culture. *Science*, 299(5603):102–105.

Waxman, S. R. and Markow, D. B. (1995). Words as invitations to form categories: Evidence from 12- to 13-month-old infants. *Cognitive Psychology*, 29(3):257–302.

Whitehouse, H. (2002). Modes of religiosity: Towards a cognitive explanation of the sociopolitical dynamics of religion. *Method & Theory in the Study of Religion*, 14(3–4):293–315.

Whiten, A., Goodall, J., McGew, W. C., Nishida, T., Reynolds, V., Sugiyama, Y., Tutin, C. E. G., Wrangham, R. W., and Boesch, C. (1999). Cultures in chimpanzees. *Nature*, 399:682–685.

Whiten, A. and Ham, R. (1992). On the nature and evolution of imitation in the animal kingdom: Reappraisal of a century of research. *Advances in the Study of Behaviour*, 21:239–83.

Whiten, A. and van Schaik, C. P. (2007). The evolution of animal 'cultures' and social intelligence. *Philosophical Transactions of the Royal Society, B — Biology*, 362(1480):603–620.

Wilensky, U. (2011). NetLogo. Center for Connected Learning and Computer-Based Modeling, Northwestern University. http://ccl.northwestern.edu/netlogo/.

Wilkinson, A., Kuenstner, K., Mueller, J., and Huber, L. (2010). Social learning in a non-social reptile (*geochelone carbonaria*). *Biology Letters*, 6(5):614–616.

Wimsatt, W. C. (2009). Memetics does not provide a useful way of understanding cultural evolution. In Ayala, F. J. and Arp, R., editors, *Contemporary Debates in Philosophy of Biology*, pages 273–292. Wiley-Blackwell, Oxford, UK.

Wolpert, D. H. (1996a). The existence of *a priori* distinctions between learning algorithms. *Neural Computation*, 8(7):1391–1420.

Wolpert, D. H. (1996b). The lack of *a priori* distinctions between learning algorithms. *Neural Computation*, 8(7):1341–1390.

Wood, M. A. (2008). *An Agent-Independent Task Learning Framework*. PhD thesis, Department of Computer Science, University of Bath, United Kingdom.

Part IV
Culture-Sensitive Technology Design

Chapter 11
Socially-Oriented Requirements Engineering: Software Engineering Meets Ethnography

Sonja Pedell, Tim Miller, Frank Vetere, Leon Sterling, and Steve Howard

11.1 Introduction

Technology can facilitate interpersonal contact in social interactions, but only if it addresses and fulfils the felt needs of people acting in their social contexts. The felt needs include those that are emotional or behavioural, such as experiencing playfulness, feeling engaged, or being capable of expressing intimacy (Howard et al. 2006; Leonardi et al. 2009; Vetere et al. 2009a; Vutborg et al. 2010; Yarosh et al. 2009). Such *socially-oriented requirements* are important to human culture but are difficult to specify and measure. Consequently, engineering systems to fulfil them is a non-trivial task. The functionality needed to facilitate a socially-oriented requirement is often unclear; for example, how do we engineer a system to ensure it is fun?

Ethnographic data can be used to inform system models and to help define socially-oriented requirements (Martin and Sommerville 2004; Viller and Sommerville 2000). However, ethnographic data does not translate into requirements in a straightforward manner. Themes extracted from ethnographic data are not functional requirements (Randall et al. 2007). Ethnographies are rich descriptions of human activities and cultural practices, and do not define the

S. Pedell (✉)
Faculty of Design, Swinburne University of Technology, 144, High Street, 3181 Prahran, VIC, Australia
e-mail: spedell@swin.edu.au

T. Miller • F. Vetere • S. Howard
Department of Computing and Information Systems, The University of Melbourne, 3010 Parkville, VIC, Australia
e-mail: tmiller@unimelb.edu.au; fv@unimelb.edu.au; showard@unimelb.edu.au

L. Sterling
Faculty of Information and Communication Technologies, Swinburne University of Technology, 3122 Hawthorn, VIC, Australia
e-mail: lsterling@swin.edu.au

V. Dignum and F. Dignum (eds.), *Perspectives on Culture and Agent-based Simulations*, Studies in the Philosophy of Sociality 3, DOI 10.1007/978-3-319-01952-9_11,

behaviour of technological systems. Ethnographic data tends to be a bottom-up view of the domain, while system models are typically derived top-down. Albeit critical, informing system models with ethnographic data remains challenging.

This chapter defines a method for addressing the gap between ethnographic data and system models created with agent-oriented techniques. We argue that the agent paradigm (Sterling and Taveter 2009) is suitable for modelling the social domain because it allows representation of the goals and motivations of agent roles and individuals. By social domain, we mean those practices that encompass cultural activities and embrace shared values. Specifically, we argue quality goals can be used to discuss socially-oriented requirements such as *having fun* and *being playful*. Our method substantiates and refines agent-oriented quality goals with attributes and new understandings about domestic cultural practices obtained from the ethnographic data.

In our method, software developers firstly define a high-level goal model that includes quality goals relevant for a specific cultural context, such as *show affection*. Ethnographic practices are then followed to obtain data about the particular domain and its value-based activities. The goal models are used as a conceptual lens through which the ethnographic data is analysed from a cultural point of view. From the data, themes are discerned, and where appropriate, a theme is attributed to a high-level quality goal. If a theme does not correspond to a quality goal, this triggers a discussion as to whether a new quality goal is required. The result is an agent model with concrete themes that exemplify how the quality goal can be fulfilled.

The domain focus of our research is the everyday life of people at home. The problem focus is how to develop technology to support interactions between family members when they are separated from each other, such as grandparents & grandchildren, intimate partners, and elderly people who are isolated from family members and friends. Thus, we are modelling human motivations and contributions to it through everyday culture. Culture here does not refer to the particular ethnic aspects that characterise it, but rather the culture of everyday interactions between family and friends. To build systems that are sensitive to this culture, we represent the everyday in terms of small, mundane yet meaningful interactions (Howard et al. 2006; Vetere et al. 2009a). Technologies for strengthening bonds within separated families must fulfil hard-to define goals such as *showing presence* and *engaging over distance*.

Our particular study examines technology for supporting the relationship between grandparents and grandchildren who are geographically separated. This study presents many interesting and challenging problems for defining innovative technologies that integrate within existing cultures. To gather field data, we use cultural probes (Gaver et al. 1999) and more specifically technology probes (Hutchinson et al. 2003) for generating insights into the interactions between grandparents and their grandchildren. The agent models are used in multiple ways throughout the development process. They serve:

- To represent an understanding of intergenerational interaction.
- As a conceptual lens through which we analyse collected field data leading to an evaluation of the original model.

- As a shared artefact between the grandparents and grandchildren, and the software engineers and ethnographers, in order to understand the cultural aspects of the home.

The approach of interleaving agent models and simple technologies helps us to improve our understanding of grandparent-grandchildren interactions and addresses the gap between ethnographic data and system models as it provides better models that are substantiated by human practice. The chapter aims to:

- Increase the modelling capability of social domains using agent-oriented concepts,
- Understand goals modelled using agent-oriented techniques, and their associated qualities, in the light of technology use over a distance, and
- Provide a method for designing and implementing non-standard quality requirements within complex social settings, such as the domestic space.

11.2 Designing for the Home

There has been a growing interest in recent years in designing domestic technologies, in particular in supporting family interactions in the home and across homes (Cao et al. 2010; Grivas et al. 2006; Judge et al. 2010; Petersen 2007; Vutborg et al. 2010; Yarosh et al. 2009; Zafiroglu et al. 2007). Domestic design has evolved from effective functional and smart technologies (Harper 2003) to more subtle and less purposeful ones, such as those for digitally mediated relationships (Grivas et al. 2006) and lightweight communication (Lindley et al. 2009b). Domestic technologies for connecting family members tend to be less concerned with informational needs, and dedicated more to connecting families in their specific and often diffuse ways, such as *passing the time* together (Howard et al. 2006). Recent research in domestic technology includes designing phatic technologies (Vetere et al. 2009b) for e.g. mediating intimacy between couples (Gibbs et al. 2005), connecting distant family members (e.g. Judge et al. 2010; Vutborg et al. 2010) and connecting older people (Cattan et al. 2005; Lindley et al. 2009a). We consider the home as a domain where culture is lived and fostered. Family life is an important carrier of our cultural life; the space where traditions are passed on. How do we develop and evaluate technologies mediating such subtle meanings such as *spending time together* that often only become apparent over the long-term or in hindsight? Currently, there is no comprehensive means of deriving culturally shaped social needs for informing agent-based quality goals.

The development of domestic technologies commonly presented in the literature is often based on field data collected in the home. Research about the contact between family members is frequently explored by introducing custom-made technology into homes and by interviewing inhabitants to investigate the success of the introduced technology (Judge et al. 2010; Lindley et al. 2009b; Vutborg et al. 2010; Yarosh et al. 2009). In most cases, little information is given on how designers progress from social needs to domestic technology, and how well the technology fulfils the family needs. Baxter and Sommerville (2010) suggest that it is not enough to analyse the situation from a socio-technical perspective and then explain the analysis to engineers. According to them approaches are needed that are pragmatic and use terminology not alien to engineers. We contribute to this work by exploring the relationship between complex value laden interactions and functional richness in domestic settings, as captured and communicated through agent-oriented models.

Getting from data of domestic lives and routines to useful and suitable technologies for the inhabitants of the home and their family relationships presents many challenges for ethnographers and software engineers. One of the big challenges for domestic design is that there is no such thing as a 'typical home' (Zafiroglu et al. 2007) or clear set tasks. Leonardi et al. (2009, p. 1703) describe the home as *"a 'territory of meaning', a place where pleasure, affect and aesthetics are deeply interwoven with the functional and utilitarian dimensions."* There is a gap between the design of domestic technologies and our understanding of the inhabitant's needs, as inhabitants represent a diverse population with non-functional and often ambiguous needs and desires that are not easy to articulate (Howard et al. 2006).

Home is a special place and designing for the home requires approaches different to traditional ones (Bell et al. 2005). In order to communicate effectively, domestically focussed design teams need a shared language, which is sensitive to their specific practices (Dearden and Rizvi 2008). Field researchers facilitating participatory design activities, and technology developers responsible for interpreting the designs for actual technologies, share the purpose of creating human-oriented technology but face very different challenges. Software engineers usually focus on future technologies and social needs are often neglected in development practice (Sommerville 2007). The ethnographers' focus is on the current lives of people. Consequently there are gaps and disconnections that both professional groups have to bridge in the design process. We aim to address this gap with the help of agent-based motivational goal models used to understand and build culturally-sensitive technologies.

In our case study, we are concerned with a particular type of social goal – the goal of *having fun*. Having fun is not simply a matter of creating a game or providing a range of communication channels. Fun is more elusive and can be subtle in its manifestations. Fun is not typically embodied in functional aims, but is expressed via social values such as simply spending time with each other. Fun comes in many forms and there are a myriad of possibilities of how fun can be realised. Research about such positive emotions around technology use is becoming increasingly important (Hassenzahl et al. 2003). Fun and enjoyment are as important in the home as productivity and efficiency are in the work context. In order to create fun-oriented domestic technologies, we need tools that are able to carry the complex, abstract and often ambiguous insights of field data collected from family cultures into the development process.

11.3 The Culture of Family Life

Other authors have focused on cultural aspects in the home and focus on the values of family members (Brown et al. 2007). The home is a space where culture is learnt, passed on and lived. Values are expressed via activities that are embedded in daily life. We are interested in exploring the social goal *having fun*. In this regard we look into the cultural aspects of one domain – the family home. We see culture as an influence on and outcome of family life.

Specifically we investigate fun resulting from intergenerational interactions and how these can be mediated. Several research projects have dealt with the grandparent-grandchild relationship and its technology support over a distance (e.g. Druin et al. 2009; Durrant et al. 2009; Lindley et al. 2009b; Vutborg et al. 2010) covering a wide range of interactions such as 'telling mobile stories' (Druin et al. 2009), 'sharing a photo frame' (Durrant et al. 2009) and 'using a lightweight messaging device' (Lindley et al. 2009b).

The grandparent-grandchild relationship is an example of a set of complex social interactions. This relationship plays an important role in our culture as the interactions between generations leads to an exchange of traditions and values. As such grandparents and grandchildren are not users in the traditional sense, but inhabitants of their particular social world with their own routines and personal lifestyles. They are living with complexities: grandparents have to fulfil a wide range of ill-defined roles to live up to being grandparents and more so being grandparents to have fun with. Interactions are based on subtle, underlying values. These values are part of the intergenerational relationship. It is a challenge to support them adequately with a suitable range of functionalities when families are separated by distance and the face-to-face exchange is reduced. Therefore, we have to look at a family's values more closely to understand emerging interactions in technology use. We analyse such interactions in the light of these values in order to draw conclusions about the nature of domestic technology. Our models help us make pressing function allocation decisions: which roles can be taken on by software agents to support culture, and which should remain with people (human agents).

11.4 Socially-Oriented Requirements Engineering with Agent Models

Typical goals in socially-oriented systems are ambiguous, non-instrumental, subtle and long term (Paay et al. 2009), and are difficult to describe and account for in ways that are appropriate for technology development. Established development tools typically deal best with clearly defined, hierarchical goals that endure over a specified time frame. Some domestic and social goals are difficult to capture with these tools. Pavon et al. (2008) argue that agent-based models are ideal for understanding the complex topics inherent in human organisations because the concepts used in these models are suitable for expressing the behavioural aspects of individuals and their interactions. We use agent modelling to represent goals and interactions, such as culturally defined roles played by the different stakeholders in the system, the related goals of the different stakeholders/roles, and the relationships between these (Sterling and Taveter 2009).

The value of matching socially-oriented studies of human interaction with user requirements elicitation methods in order to abstract activity and embed technology into social contexts has been acknowledged (e.g. Viller and Sommerville 2000). Other researchers describe bridging the gap between the output of field studies and the required input to system development through meta-modelling (Iqbal et al. 2005). However this is not straightforward for socially-oriented requirements. Eliciting socially-oriented requirements from field data involves working in a milieu in which it is essential to capture concepts accurately but flexibly and at a high level, without losing the liveliness and vitality of those concepts through overly detailed specification. For example, the role of a grandparent does not come with an established list of responsibilities or a minimum performance plan. The roles that grandparents see themselves playing in the life of their grandchildren are highly dependent on many factors such as the individual, their experience with their own grandparents, and their cultural background.

In our approach we use the following main components:

1. Starting with motivational models with a focus on quality goals.
2. Implementing lean, but focused technologies.
3. Lightweight evaluation of quality goals using ethnographic studies.
4. Substantiating quality attributes of use activities in quality clouds.
5. Iterative exploration and discussion of social requirements.
6. Refining of user needs for domestic technologies.

In this paper, we present the components of the method and not the overall process. The activities of these components take place iteratively, depending on the available knowledge of the user domain.

11.4.1 Motivation Models

Here we build further on the work of Sterling and Taveter (2009). Their research has focused on how to make high-level agent-oriented models palatable in design discussions. This is achieved by using goal models that have a straightforward and easy syntax and semantics. Goal models are useful at early stages of requirements analysis to arrive at a shared understanding (Guizzardi et al. 2005; Jureta and Faulkner 2007); and the agent metaphor is useful as it is able to represent the concepts that we want to capture for socially-oriented systems, such as agents (people) taking on roles associated with goals. These goals include quality attributes that are represented in a high-level pictorial form and that are used to inform and gather input from stakeholders. In Sterling and Taveter's notation, goals are represented as

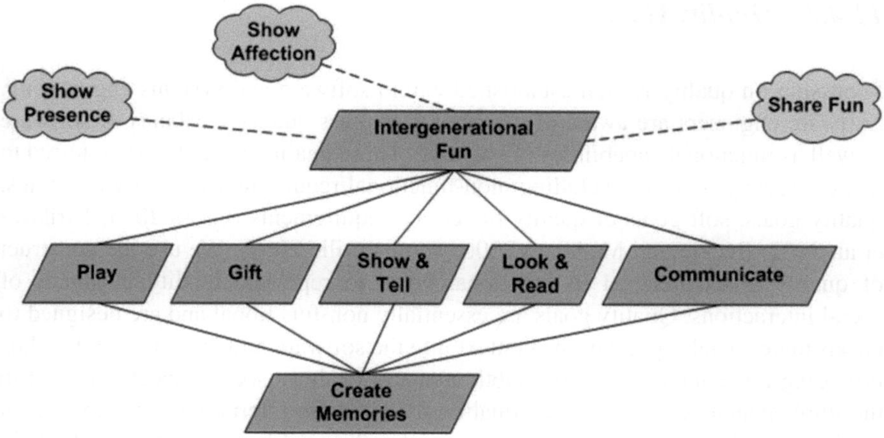

Fig. 11.1 Agent-oriented model representing *intergenerational fun*

parallelograms, quality goals are clouds, and roles are stick figures. These constructs are connected using arcs, which indicate relationships between them.

We started with a set of high-level qualities or values of grandparent-grandchild interactions: *share fun*, *show presence*, and *show affection* (component 1). Activities that would support these specific aspects of intergenerational interactions are e.g. *gifting*, *playing*, and *show & tell*. These values and abstract activities are represented in a high-level goal model (see Fig. 11.1). Both values and activities were derived from the results of former research that we conducted in this domain (Paay et al. 2009; Vetere et al. 2009a). The goals depicted as parallelograms represent meaningful activities in the grandparent-grandchild interaction. The quality goals represented by clouds are high-level attributes that are often subjective, context-specific, and imprecise, and are therefore able to express the nature of intergenerational fun. We included such quality goals as part of the design discussions since they accounted for social values embedded in intergenerational fun interactions. The agent-oriented model of *Intergenerational Fun* depicted in Fig. 11.1 served as a common basis for design discussions of building three technologies by three different design teams – one of the technologies discussed in detail in the next section.

Quality requirements at the early stages of elicitation tend to be imprecise, subjective, idealistic and context-specific (Jureta and Faulkner 2007). Garcia and Medinilla (2007) describe high-level quality goals as a specific form of uncertainty that can be used as a descriptive complexity reduction mechanism and to model and discuss uncertainties in the environment. In our requirements elicitation process, we seek complexity reduction without losing the richness of the social concepts themselves. Instead of eliminating uncertainty early in the process, we embrace it and withhold design commitment, at least until there is clarity and understanding (Gause 2000). High-level goals associated with activities can act as a point of reference for discussing the usefulness of design alternatives to achieve these goals instead of decomposition into single requirements.

11.4.2 Quality Goals

Focussing on quality is well established within software and systems engineering. Software engineers are aware of the need to express quality attributes of software as well as functional capabilities of software. These quality attributes are referred to using a variety of terms including: non-functional requirements, quality attributes, quality goals, soft goals or quality of service requirements (Gross 2005; Kirikova et al. 2002; Reekie and McAdam 2006; Sommerville 2007). We use the construct of quality goals attached to functional goals to represent quality attributes of social interactions. Quality goals are essentially non-functional and are designed to encapsulate social aspects of the context into the software requirements model, thus providing a mechanism to carry subtle nuances of those social aspects through to the implementation phase. These quality goals remain interpretably flexible, even until the final product, opening up a variety of possible interpretations both in the design and use of the system. Some of the goals might influence the choice of functions and some might remain with the human agents – in our study, the grandparents and grandchildren. There is benefit in articulating socially-oriented quality goals without the need to resolve them into measurable goals or requirements that are implementable. We introduce culturally-influenced qualities as part of a socio-technical system tied to the motivations of their users. Sterling and Taveter's agent-oriented models allow the expression of non-functional requirements by attaching quality goals to goal models (Sterling and Taveter 2009). In our approach there is a direct pairing between system goals and quality goals, whereas non-functional goals do not generally have a direct relationship with functional goals (Chung et al. 2000). This makes it more difficult to carry them through the process in an unresolved state. Relating an abstract and unresolved quality attribute to a system goal enables a focus on social goals throughout the design process.

From a software engineering point of view, the models enable us to take the outputs from a field study and use them to inform system development. This is achieved by taking account of the richness of human social interaction provided by the probe data, encapsulating quality attributes of that interaction into quality goals in the models, and using these models as inputs to the design process.

11.5 Using Technology Probes to Obtain Cultural Data

In our case study we built three technology probes that were inspired by the motivational model from Fig. 11.1; *Collage, electronic Magic Box,* and *Storytelling*. The technologies were seen as instances of the goal model and emphasise various goals of the model. While Collage has its focus on 'playing' (Vetere et al. 2009a) and Storytelling on 'show and tell' activities (Vutborg et al. 2010), the electronic Magic Box has its focus on 'gifting'. Next we describe the results and the procedure for informing quality goals focussing on the electronic Magic Box.

11.5.1 Benefits of Probes in Developing Domestic Technologies

Probes (Gaver et al. 1999) are particularly suited to investigating people's everyday life in situations difficult to reach with traditional social science methods, such as questionnaires, interviews, focus groups or participant-observation. Rather than relying on the presence and intervention of the researcher, probes are designed to encourage and empower subjects to collect data themselves (Arnold 2004). The participants use the probes to provide some insight, at their discretion, about their daily lives. Probes are specifically suitable for collecting data in the domestic domain through their ability to capture the nuanced aspects of everyday life (Arnold 2004; Hemmings et al. 2002). Information and story generation are two important benefits that we see in the use of probes. Our approach required minimal ongoing intervention from the researchers, while allowing observation of the transactions between the participants. Therefore, the three technologies had logging capabilities to monitor and record the use of the applications serving as technology probes (Hutchinson et al. 2003). The interactions of all three systems, the messages and photographs were saved on several servers.

11.5.2 Benefits of Goal Models

The agent-oriented models are particularly suitable to be combined with technology probes in field studies. Firstly, we see agent-oriented models as a suitable way to express field data. As data gathered using probes are intentionally fragmentary and unstructured, the process of translation from field data to the abstract generalisation required in development is problematic. A process of combining technology probe data collection and agent-oriented models allows us to talk about intangible outcomes; such as that arising from fieldwork which can be surprising, complex, but subtle. The agent-oriented models provide a place where abstract design concepts can be collected and represented (Pedell et al. 2009). They are a lens through which use activities can be analysed and recorded and then discussed among researchers and software engineers. Secondly, agent-oriented models are part of a development methodology and can be combined with motivational scenarios, roles and domain models (Sterling and Taveter 2009), each of them describing and providing context of the domain, which is important because contextual information offered by technology probes is often lost after data analysis.

11.5.3 Technology Set Up

The technology probes (component 2) synchronous 17 inch touch screen monitors (as this study was pre iPAd it is worth to point out) for display, and mobile camera phones for sending photographs and messages that were shared amongst the grandchildren and grandparents households. Each household was allocated one

Fig. 11.2 Electronic Magic Box. A treasure box filled with a message (**a**) doing the maze (**b**)

mobile phone and one touch screen; that is the grandparent household shared one phone and one touch screen between them and the grandchildren and parents shared one phone and one touch screen between them.

The mobile phones were important as we wanted sharing of everyday experiences. Grandparents and grandchildren could carry the phones with them and share photographs of events and ideas with the others sending it to the system without the need to be home. We chose technologies that constrained use as little as possible thereby facilitating flexible interactions without strict assumptions about how the technology was to be used. The systems were placed in high traffic areas in the family homes (e.g. lounge room or kitchen). While our focus was on the grandparent-grandchild relationship, the parents took on an important role in facilitating interactions and observing them without being active users.

The electronic Magic Box, the technology probe described here in detail allowed the sending of a treasure box that could be filled with photographs and messages. Figure 11.2a shows the layout of the opening screen. On the left side, seven picture-based links (home, magic box, scroll, collection book, settings, admin, and logout) can be found that guide the user to a number of destinations within the application. The box is placed in a forest of fern trees and appears either closed (a new box has arrived) or open (no new box has arrived). A scroll either sealed or with a broken seal indicates if the box in the other household has been opened and the content been looked at. In order to be able to access the content the receiver has to play a maze game (Fig. 11.2b) to 'find' and open the box. An opened message can be saved in a collection book. Emphasis in this application was placed on the goal *gifting*, but the concept certainly carried elements of and was inspired by the other high-level goals of *playing*, *show & tell*, *look & read*, *communicating* and *creating memories*.

11.5.4 Study Design and Participants

We introduced the three probe technologies to three families. The applications were installed in the family homes for 3–6 weeks over an period of 4 months (component

3). The children were aged between 5 and 9 years and the grandparents lived between 8 and 16 km away from the children. All grandparents had regular contact with their grandchildren and all described having a strong and loving relationship.

11.5.5 Data Collected

We conducted three to four interviews per household about the probe use (usually grandparent and parent/grandchild interviews were conducted separately) – in total 20 interviews were conducted. The parents' presence in the grandchildren interviews was an important source of information as they were observing the ongoing interactions without being active users and were able to comment on changes since the introduction of the technology probes. During the interviews we did not specifically ask questions about *playing* or *gifting*. These goals are implicit in the system as we tried to provide activities that supported these goals. We were more interested in the social interactions and how the qualities were judged by the participants. For example we would ask: "what kind of interactions did the system support?" and "what activities did you particularly enjoy?". If we did not obtain feedback that using the technology was fun, then we would have felt our original model was invalidated. The technology probe data collected with the electronic Magic Box consisted of 102 boxes (electronic letters and photographs), meta data about each box such as send times, and data from seven interviews.

11.6 Analysing and Discussing Cultural Data

The success of a design in achieving its goals can really only be investigated after implementation. The technology probes embodied certain goals of the goal model. We purposely kept the goals at a high level that was representative and comprehensive, as determined by the development team, but independent of any future implementation. Therefore, we were able to link the qualities learned during the ethnographic studies to the motivational models. The transcribed interviews together with photographs and electronic letters were analysed using content analysis (Patton 2002). The quality goals played the role of overarching themes for analysis. We explored intergenerational activities and interactions rather than technology per se. We were able to find sub-themes for all of the quality goals and therefore to learn more about each goal in the light of typical social activities between grandparents and grandchildren (component 4). Each sub-theme was briefly described and substantiated by compelling examples and instances of these goals in the context of intergenerational fun during use.

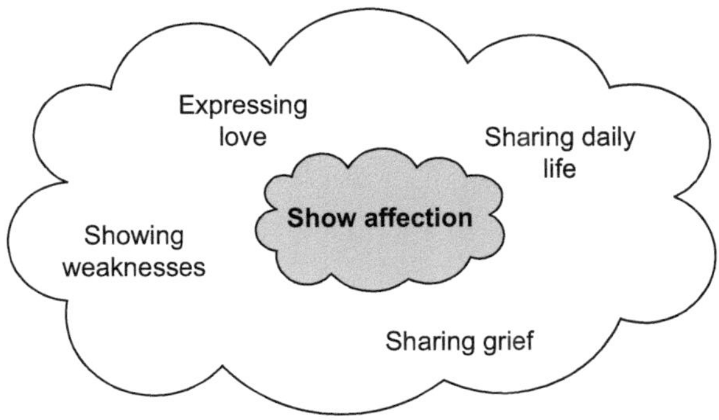

Fig. 11.3 Resulting quality cloud of analysed quality goal *show affection*

We analysed the interview data according to what we could learn about the quality goals, using the model in Fig. 11.1 as an interpretive lens. The photographs and messages were analysed and discussed at regular project meetings. The essence of the quality goals was based on experiences and judgement of the participants regarding their interactions. Therefore, the interview data played a major role in this analysis as we wanted to expand from the activities and original goals to inform the quality goals. The photographs and messages were used mostly to confirm and illustrate the results with particular episodes and participant stories. The analysis helped us to keep the focus on the human needs with the technology as mediator. The purpose of the technology was not just to support intergenerational fun, but to support the development team in further investigating the qualities of the social goals in the model themselves.

The sub-themes that emerged from our data analysis were organised as characteristics to the quality goals into so-called *quality clouds* (Fig. 11.3). The quality clouds consist of one quality goal – still linked to a functional goal – with associated qualities factored around. The quality clouds can be seen as an abstract representation of field data into which we are able to zoom into the associated quality goal more closely. The quality cloud shown in Fig. 11.3 concerns the quality goal *show affection* and its associated quality attributes. In this process the sub-qualities or quality attributes were formulated into adjectives to re-connect the qualities in discussions more easily to the functional goals they are attached to.

Each sub-quality of a main quality goal is briefly described and directly linked to the respective quotations in the interview data. In that regard the quality goals were augmented by ethnographic data. While we were interested to group the sub-qualities to our existing quality goals, in order to substantiate them with our field data, we permitted new main quality goals to emerge from the ethnographic data, and hence allow changes to our overall goal model. As part of the method, in the event important activities or themes evolve for which we cannot find a home, we define new quality goals.

11.7 Results

11.7.1 Substantiating Quality Goals: Show Affection

Below we present a sample of the interview data that demonstrates the process for elaborating the quality cloud *show affection*. Some of the sub-qualities from the clouds that brought us unexpected insights are described in more detail.

Share daily life: The aim was to share with and include the other family member in the happenings of the own life.

> Andrea tells nanna everything! For example the photo of the 'Dog sick'! – anything that took her interest on that particular day. That particular hour – absolutely everything she wants to tell her nanna [parent].

Show weaknesses: Family members were comfortable not only showing their best side, but also failures and weak points, because there is a loving trust within the relationship.

A challenge for most of the grandparents was managing the technology. Uncovering this kind of "weakness" is a very intimate act in itself. Problems dealing with the electronic Magic Box were often communicated in a humorous way or loaded with self-irony making the technology handling a shared episode in itself. A nice example was one grandmother sending a picture of her granddaughter along with this message:

> *Dear Andrea, in trying to send this photo to you I burnt my steak I am having for dinner, yuk!!!* After this she took a photo of her burned frying pan as well and sent it (Fig. 11.4a): *When I tried to send this message Thursday the machine told me to try again, so here I am. This is the pan I burned while trying to enter the project!!!!!!!*

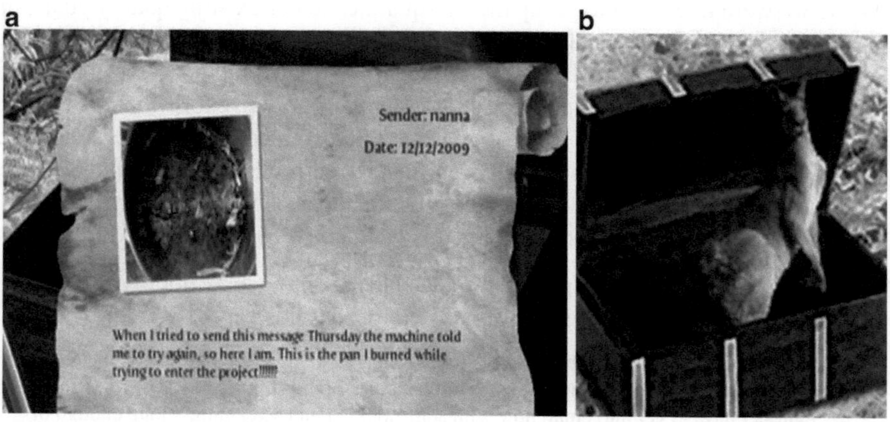

Fig. 11.4 The burnt steak (**a**) and the surprise kangaroo (**b**)

Our families tended to show themselves to people they trust and love. This grandmother assured the researchers that the pan was 'all clean again' and that she had no more disasters. In a similar way one grandmother sent a photograph of her messy desk.

> *"This is my messy desk. I am trying to catch up with office work"*. The granddaughter took it up immediately as something funny and kept saying in the interview: *"Granny you are messy as well – you sent me this photograph of your desk."*

That the grandparents admitted to weaknesses being adults and "should know better" was received as something very special by the grandchildren.

Express love: It was very common to send a message that explicitly expressed love or physical closeness.

> *"I love you"* – messages and the building of a little sculpture with a sign *"Nana gives the best hugs"* sent as photograph are examples for the mutual felt love.

Share grief: The electronic Magic Box was particularly well suited in mediated shared emotions. There was sometimes an urge to transfer something important and emotional. One example was when the granddaughter's dog got really sick and died.

> The granddaughter wrote her Nanna accompanied with a really sad picture of herself: *"I really miss Sam – really really!"* Her granny answered: *"I have been thinking of her too, but she was very sick & you wouldn't want her to suffer, would you?"*

Overall, the electronic Magic Box mediated comfort and shared understanding, in addition the exchange of fun messages.

11.7.2 New Quality Goals

While we were interested to group the sub-qualities to our existing quality goals, in order to substantiate the quality goals with our field data, we permitted new main quality goals to emerge, and hence allow changes to our overall goal model. As part of the method, in the event important activities or themes evolve for which we cannot find a home, we define new quality goals. Qualities emerging that we could not group with our existing quality goals were themes surrounding the technology use itself still being closely connected to positive feelings – often explicitly described as fun. The new quality goal that emerged is *build confidence*, shown in Fig. 11.5.

Learning: One important aspect was being able to continuously improve managing the technology. Some of the grandparents expressed it this way:

> It is quite interesting to see where we started: "I didn't find a photo, but here is the text". Next time I was able to send the text as well. It is a bit of fun [GP]
>
> I guess I have to get into email now with some kicking and screaming – I am enough of a dinosaur. I think I am ready [GP].

When an empty box was sent: a kangaroo would jump out of the box. The families described they had a lot of fun when this function was discovered. One grandchild could not figure out how this had happened and could not get enough of

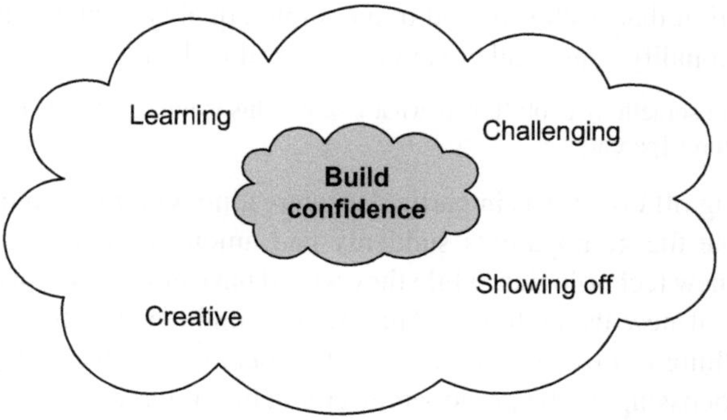

Fig. 11.5 New quality goal *build confidence*

the kangaroo magically jumping out of the box (see Fig. 11.5b); the grandmother was proud that she was able to do something unexpected with the technology that links in with building up confidence.

Creativity: It was seen as a challenge, but worthwhile effort to get the material for a creative message together.

> When I had an inspiration and took a photo then that was a bit of fun [GP].

Mastering a challenge: Sending a box off every day is an activity that is achievable, but still takes some effort such as to 'take a photo', 'pack it' in the box and 'send off'.

> Magic box is sort of you are in Prep and then you are in grade one. When you start school you go up to grade one. Collage [the first technology probe we had introduced to that family] was Prep and electronic Magic Box is like grade one – a step up. There is a little bit to it – to go into it. We graduated [GP].
>
> I am sorry to see it go besides the fact that I don't have to wake up and think what should I send today? [GP].

All grandparents did master the challenge after sending a few boxes. The children did not have any problems with the use of the electronic Magic Box. However, the maze was not only fun for the children – here was where the children got their challenge as the technology was for them in most parts no problem to handle. Animal buttons for settings from easy (koala bear) to wicked (frog) were indicators for improvement. This challenge was important to the children without being competitive with others. It would engage them indirectly in an interaction with grandparents as they could play this game only when a message was sent by the grandparents and therefore was associated with the grandparents' interactions. In this regard the maze also fulfilled a balancing function for the fact that the children did not feel challenged by the technology. The quality goals are fulfilled for all users even though they can take different forms.

Showing off: Showing the application to people like neighbours, friends and other family members with a feeling of pride. This theme is a clear sign that confidence indeed had been built up and another example or measure for validating the success of the application that is closely tied to a complex quality goal and not to a certain piece of functionality. One mother said about her daughter:

> Showing them something cool: 'this is what I've got this is mine' – this is my phone and I can send pictures [parent].

The showing off effect was in particular interesting with the grandparents. There was a new role the grandparents suddenly had among their peers. They became advocates for new technologies, while they would have never anticipated themselves as champions of new technologies. They found confidence in the technology that really had nothing to do with anything we had planned the technology to provide. However, in focussing on the grandparent-grandchild interaction and in keeping the technology lean and simple until the quality goals were better understood, we had catered for enjoyable use in a very substantial way (component 5).

11.7.3 Understanding Socially-Oriented Requirements

Figures 11.3 and 11.5 show the substantiated quality goals after the ethnography was done (component 6). Newly formulated requirements are to a large extent influenced by the new quality cloud *build confidence*. Building confidence is part of the intergenerational interaction and it has implications on how the technology should be designed: we learnt not to put everything in an application at once, because it scares the grandparents away. We now maintain simple screen views and a layered application instead of an application packed with functionality. Aiming for simplicity is not only based on the lack of confidence of many grandparents to deal with complex technology, but is suggested in the nature of strong-tie relationships themselves. It is apparent in the sub-themes that these technologies rely on an existing rich and loving relationship. To support the long-term interactions between grandparents and grandchildren, technologies need to mediate these subtle but complex relationships within the family context and routines. We cannot evaluate the success of the technology per se. We have to evaluate if the use of the technology supports the quality goals of the goal model. We use the model to see if it is indeed a representation of the socio-technical system (technology use of the grandparents with their grandchildren for the purpose of having fun).

Another important insight was "the other side of fun". Certain familial values tend to be marginalised. For example, disclosing weaknesses or failure – and laughing about them – or the demonstration of grief and openly dealing with it, are not normally identified as laudable values in systems design. In our study, the grandmother does not try to brush the grief away with some happy comment, but she honestly acknowledges that the loss of the loved dog indeed is sad. Dealing with these kinds of emotions is just as important for a strong-tie relationship as demonstrating love, play together and laugh about a joke. It is no contradiction that technologies for intergenerational fun also allow and even aim for activities that deal with aspects we would normally avoid to show openly or associate with fun.

11.7.4 Summary of Method for Modelling Social Interactions

We have presented a method for substantiating quality goals in the development of domestic technologies to support interactions between grandparents and grandchildren. We used agent-based models for representing the goals and motivations of individuals with a focus on family values. We described six components as part of our approach. The components have allowed us to explore the trade-offs between functional richness and use of technology in the home. We use agent-oriented models to record the high-level goals and their quality attributes to represent social interactions, which can provide an account for social concepts such as fun or intimacy. The agent models proved to be particularly suitable to express culturally sensitive data obtained from field studies. As data gathered using probes is fragmented and unstructured, the process of translation from field data to the abstract generalisation required in development is difficult. The models provide a place where abstract design concepts can be represented, helping the researchers and software engineers come to a shared understanding of the social domain.

11.8 Conclusions

In this chapter we have explored the use of agent-oriented models during system development in order to elicit, understand, and represent socio-cultural aspects of everyday life. In our case the domain of interest has been domestic technology use. A process for combining ethnographic data and agent-oriented models informs and substantiates understandings of the domain, family values and activities. The process generates findings that are often surprising, complex, and nuanced. The agent-oriented models provide a place where abstract social activities and qualities can be collected and represented. They are a lens through which use activities can be analysed and recorded and then discussed amongst researchers and software engineers. A number of benefits emerged from our approach:

Sharing and making explicit. Fun, as many other social concepts, has many facets and it is beneficial to agree on a high-level view when building a socio-technical system. A shared view between software engineers and ethnographers helps to orient communication and focus the team on the relevant data during collection and analysis. There remains the ongoing possibility to change and refine this shared view during discussions. The motivational goal model allowed us to discover new quality goals whilst we learnt about new social aspects and attributes of the initial ones.

Grounding design in data. During development we were able to keep associations between the ethnographic data and the motivational models. We could understand the qualities of the clouds as examples of real social activities. These examples and associations were meaningful during the software engineering process when discussing high-level requirements for building new domestic technologies. With the agent-oriented models we were able to see the users' motivations made real in design.

Closing the gap and interleaving processes. There is a strong interplay and information exchange between the field data and the agent-oriented models. The standard software engineering process is a top down process. We used the high-level structured view –the quality goals – as a lens to analyse field data. Importantly, the fluid process influences the bottom-up information flow also. We changed the models as we discovered new qualities and learnt more about existing quality goals. In this sense we were matching two different perspectives, top-down and bottom-up. The suitability of our technologies demonstrates the extent to which the gap was closed and where we still had to achieve a better match between initial understanding and consequential implementation of the cultural model of intergenerational fun.

Traceability of motivations. This method is repeatable and traceable, as evidenced in statements such as "I made this decision because this was fun for so many families when 'x' was happening". The possibility to refer to the context and trace motivations is a crucial process in the development of socio-technical systems.

Validation. With the quality clouds we were creating a set of new testing artefacts. They were useful in the process to validate associations between activities and high-level goals and evaluate the degree of the match between the two. This took place in a participatory manner – including the grandparents and the grandchildren.

Overall the approach described above assisted ethnographers and software engineers in arriving at a shared understanding of social goals and the related interactions in a way that became useful in ongoing software development for the social domain.

Acknowledgments We would like to thank the participating families. This research was funded by an Australian Research Council Discovery Grant (#DP0880810).

References

Arnold, M. The connected home: probing the effects and affects of domesticated ICTs. In Proc. PDC 2004, ACM Press, 1–4. (2004)

Baxter G., and Sommerville, I. Socio-technical systems: From design methods to systems engineering. Interacting with Computers. 23, 4–17. (2010)

Bell, G., Blythe, M., and Sengers, P. Making by making strange: Defamiliarization and the design of domestic technologies. ACM Trans. Comput.-Hum. Interact. 12, 2, 149–173. (2005)

Brown, B., Taylor A., Izadi S., Sellen A., Kaye J. J., and Eardley R. Locating Family Values: A Field Trial of the Whereabouts Clock. In Proc. UbiComp'07. 354–371 (2007)

Cao, X., Sellen, A., Brush, A. J. et al. Understanding family communication across time zones. In Proc. of CSCW 2010, ACM Press, 155–158. (2010)

Cattan, M., White, M., Bond, J., & Learmouth, A. Preventing social isolation and loneliness among older people. Ageing & Society, 25, 1, 41–67. (2005)

Chung LK, Nixon BA, Yu E, Mylopoulos J. Non-Functional Requirements in Software Engineering. Kluwer Publishing. (2000)

Dearden, A., and Rizvi, H. Participatory IT Design and Participatory Development: A Comparative Review. In Proc. PDC 2008, 1–10 (2008)

Druin, A., Bederson, B. B., and Quinn, A. Designing intergenerational mobile storytelling. In Proc. IDC 2009, ACM Press, 325–328. (2009)

Durrant, A., Taylor, A. S., Frohlich, D., Sellen, A., and Uzzell, D. Photo displays and intergenerational relationships in the family home. In Proc. BCS HCI 2009, ACM Press, 10–19. (2009)

Garcia A. and N. Medinilla. The ambiguity criterion in software design. In International Workshop on Living with Uncertainties (IWLU'07). ACM, (2007)

Gause. D. User DRIVEN design the luxury that has become a necessity, a workshop in full lifecycle requirements management. In ICRE 2000, Tutorial T7, (2000)

Gaver B., Dunne T., and Pacenti E. Design: Cultural Probes. Interaction 6(1): 21–29. (1999)

Gibbs, M. Vetere F., Bunyan M., and Howard S. SynchroMate: A Phatic Technology for Mediating Intimacy. In Proc. DUX 2005, AIGA 37, 5 pages (2005)

Grivas, K. Digital Selves: Devices for intimate communications between homes. Personal Ubiquitous Comput. 10, 2–3, 66–76. (2006)

Gross H. Component-based Software Testing with UML. Springer. (2005)

Guizzardi R. and A. Perini. Analyzing requirements of knowledge management systems with the support of agent organizations. Journal of the Brazilian Computer Society (JBCS)-Special Issue on Agents Organizations, 11(1):51–62, (2005)

Harper, R. Inside the Smart Home: Ideas, possibilities, methods. In Harper, R. (Ed). Inside the Smart Home: Interdisciplinary perspectives on the design and shaping of domestic computing. Berlin Springer, 1–13. (2003)

Hassenzahl, M. The thing and I: Understanding the relationship between user and product. In M. Blythe, C. Overbeeke, A. Monk, P. Wright (eds.). Funology: From usability to enjoyment. Kluwer, 31–42. (2003)

Hemmings, T., Clarke, K., Crabtree, A., Rodden, T. and Rouncefield, M. Probing the probes. Domestic Probes and the Design Process. Proc. PDC 2002, 42–50. (2002)

Howard, S., Kjeldskov, J., and Skov, M. (eds). Pervasive Computing in the Domestic Space. J. Personal and Ubiquitous Computing, 329–414. (2006)

Hutchinson, H., Mackay, W., Westerlund, B., et al. Technology Probes: Inspiring Design for and with Families. In Proc. CHI 2003, ACM Press, 17–24. (2003)

Iqbal, R., James, J., and Gatward, R. Designing with ethnography: An integrative approach to CSCW design. Advanced Engineering Informatics, 19:81–92. (2005)

Judge, T. K., Neustaedter, C., and Kurtz, A. F. The family window: the design and evaluation of a domestic media space. In Proc. CHI 2010, ACM Press, 2361–2370. (2010)

Jureta I. J., and Faulkner S. Clarifying goal model. Conceptual modelling. Proc. ER, 139–144. (2007)

Kirikova M, Grundspenkis J, Wojtkowski W, Zupancic J, and Wrycza S. Information Systems Development: Advances in Methodologies, Components, and Management. Springer. (2002)

Leonardi, C., Mennecozzi, C., Not, E., et al. Knocking on elders' door: investigating the functional and emotional geography of their domestic space. In Proc. CHI 2009, ACM Press, 1703–1712. (2009)

Lindley, S. E., Harper, R., and Sellen, A. Desiring to be in touch in a changing communications landscape: Attitudes of older adults. In Proc. SIGCHI 2009, ACM Press, 1693–1702 (2009a)

Lindley, S. E., Sellen, A., and Harper R. Bridging the gap between grandparents and teenagers. CHI 2009b http://research.microsoft.com/en-us/groups/sds/ publications.aspx

Martin D, Sommerville I. Patterns of Cooperative Interaction: Linking Ethnomethodology and Design. ACM Transactions on Computer-Human Interaction 11(1):59–89. (2004)

Paay, J., Sterling, L., Vetere, F., Howard, S., and Boettcher, A. Engineering the Social: The Role of Shared Artifacts, IJHCS, 67(5):437–454. (2009)

Patton, M. Qualitative Research and Evaluation Methods. Sage Publications, Thousand Oaks California, (2002)

Pavon, J., Arroyo M., Hassan, S., and Sansores. C. Agent-based modelling and simulation for the analysis of social patterns. Pattern Recognition Letters, 29(8):1039–1048, (2008)

Pedell, S., Miller T., Vetere, F. et al. Having fun at home: interleaving fieldwork and goal models, Proc. OZCHI 2009, 309–312. (2009)

Petersen, M. G. Squeeze: designing for playful experiences among co-located people in homes. In Proc. CHI 2007, ACM Press, 2609–2614. (2007)

Randall, D., Harper, R., and Rouncefield, M. Fieldwork for Design: Theory and Practice. Springer. (2007)

Reekie H., and McAdam R. A Software Architecture Primer: Angophora Press. (2006)

Sommerville, I. Software Engineering (8th ed). Addison Wesley, Harlow, Essex. (2007)

Sterling, L., and Taveter, K. The Art of Agent-Oriented Modelling. MIT Press. (2009)

Vetere, F., Davis, H., Gibbs, M., Howard, S. The Magic Box and collage: responding to the challenge of distributed intergenerational play. International Journal of Human Computer Studies 67 (2), 165–178. (2009)

Vetere, F., Smith, J. and Gibbs, M. Phatic Interactions: Being Aware and Feeling Connected. In P. Markopoulos, B. de Ruyter, and W. Mackay (eds) Awareness Systems: Advances in Theory, Methodology and Design. Springer Verlag, London, 173–186. (2009)

Viller, S., and Sommerville, I. Ethnographically informed analysis for software engineers. IJHCS, 53(1):169–196. (2000)

Vutborg R., Kjeldskov J., Vetere F., and Pedell S. Family Storytelling for Grandparents and Grandchildren living apart. In Proc. NordiCHI 2010, ACM, 531–540. (2010)

Yarosh, S., Cuzzort, S., Müller, H. and Abowd, G. D. Developing a media space for remote synchronous parent–child interaction. In Proc. IDC 2009, ACM Press, 97–105. (2009)

Zafiroglu, A. and Chang, M. Digital homes on wheels: designing for the unimagined home. Personal Ubiquitous Comput. 11, 5, 395–402. (2007)

Chapter 12
Cultural Broker Agents: A Framework for Managing Cultural Misunderstandings

O. González, J.-P. Barthés, and F. Ramos

12.1 Introduction

As an increasing tendency, advances in technology and communications, global partnerships, and expansion of transnational enterprises drive the creation of distributed and multicultural work teams. Such teams highly depend on clear and efficient interpersonal interactions and in a strategy for avoiding misunderstandings deriving from cultural differences.

In such multicultural scenarios, it would be desirable to have a framework to: (i) formalize and measure the culture of users; (ii) define cultural misunderstandings prone to occur; (iii) define proper reactions for each type of cultural conflict; and (iv) timely show explanations and advises to users when potential cultural misunderstandings are detected in their interactions.

This chapter proposes a framework with such capabilities, namely a multi-agent system (MAS), where computer supported interactions are managed by cognitive agents with cultural capabilities called cultural broker agents (CBA). Under this approach, each user is supported by a CBA which formalizes his actions. When users interact, the corresponding CBAs act as intermediaries, capturing information, exchanging messages, and detecting potential cultural misunderstandings based on cultural differences and patterns in interactions.

O. González (✉)
CINVESTAV Unidad Guadalajara
e-mail: oglez@gdl.cinvestav.mx

J.-P. Barthés
Université de Technologie de Compiègne BP 529 F-60205 COMPIEGNE France
e-mail: Jean-Paul.Barthes@utc.fr

F. Ramos
Cinvestav Unidad Guadalajara, Mexico
e-mail: framos@gdl.cinvestav.mx

V. Dignum and F. Dignum (eds.), *Perspectives on Culture and Agent-based Simulations*, Studies in the Philosophy of Sociality 3, DOI 10.1007/978-3-319-01952-9_12,

Cultural preferences of users are formalized by CBAs as cultural profiles inspired in the theory proposed by Hofstede (1996), who summarizes the culture of individuals as a set of quantitative features.

CBAs formalize interactions by capturing the intention behind each user action. Each time a user interacts with another, the corresponding CBA obtains the illocutionary force, i.e. the intention of the action (Searle 1975). This approach is based on speech act theory (Austin 1975) and the language-action perspective (Winograd and Flores 1987).

Cultural conflicts are defined in different stages of interactions as conditions involving differences in cultural profiles and additional information which vary according to the conflict being defined.

12.2 Models of Culture

Culture is a complex concept with many definitions; none of them completely compatible (Paez et al. 2003). However, most anthropologists agree defining culture as a set of socially acquired patterns for thinking, feeling and acting.

Several researches are aimed at developing measurable models of culture; the most common approach is to decompose culture in a set of cultural dimensions representing the degree of acceptance or rejection for specific behaviors (Hofstede 1996). Depending on the scope of each research, different models base on diverse aspects like: attitudes when interacting with people (Hofstede 1996; Trompenaars 1998), attitudes towards time (Trompenaars 1998; Hall 1989) and the environment (Trompenaars 1998), communication (Hall 1999), corporal language (Hall 1990), or biological needs (Schwartz 2003).

Such models are commonly used in technological developments which require a method for measuring culture. Which model to select depends on the requirements of the application; examples are communication tools (Dutsbar and Hofstede 1999), virtual reality simulations (Llobera et al. 2010), computer supported collaborative work (Kamel and Davison 1998), and design of user interfaces (De Troyer et al. 2006).

We found the model of Hofstede (1996) the most adequate for developing our framework for intercultural collaboration. Hofstede's model is derived from studies on organizational culture, the collection of values and norms that control the way in which people interact in organizations (Hill and Jones 2006).

12.2.1 Hofstede's Cultural Dimensions

Based on data collected from IBM workers in 70 countries, Hofstede identified 4 cultural dimensions affecting interactions in organizations (Hofstede 1996):

- Individualism vs. Collectivism. Degree in which people prefer acting by themselves, or alternatively, as members of a group.

- Power Distance. Degree in which it is accepted in a society that people is treated differently according to social or economical position. A low value reflects a shared belief that people must be treated in the same way.
- Masculinity vs. Feminity. Extent to which a society tends to be more interested in material possessions than in personal relationships and quality of life. Masculine societies value competitiveness, assertiveness, and ambition; feminine societies value family, friendship and quality of life.
- Uncertainty Avoidance. Degree in which a society accepts uncertainty and risk. Societies with high values avoid changing processes and make efforts for reducing ambiguity as much as possible.

In the mid 1980s, Michael Bond (1988) noticed differences of thinking between Eastern and Western cultures and developed a questionnaire about how time is managed when acting and fixing objectives. The questionnaire was applied in 23 countries and the analysis of results led to the addition of a fifth dimension to the model:

- Long Term Orientation vs. Short Term Orientation. How important is in a society the future compared to the importance given to past and present. On the long term oriented pole, people value persistence, perseverance, thrift, and planning. Short term oriented societies give special value to stability, traditions and fulfillment of obligations.

Based on an analysis of the answers of its questionnaire, Hofstede computed values for his cultural dimensions describing stereotypes for different countries. Most of the criticisms to Hofstede are focused on the sample he used for obtaining such values. McSweeney (2002) argues that considering groups as big as national populations sharing a common culture is not a valid hypothesis; he also remarks that answers were provided by respondents with similar demographical information, neglecting the diversity of subcultures inside a country. Bryman (1988) states that the number of respondents is not representative for some countries; finally, values obtained by Hofstede are old and ignore global changes of last years.

Despite these criticisms, we consider the theoretical framework, i.e. the proposed dimensions for abstracting culture, represent an adequate model for the scope of our work. According to this, we obtained new values for cultural dimensions for some countries. Like Hofstede, we compute national cultural profiles statistically based on questionnaires; however, we applied them to social network users which provides some remarkable advantages: (i) given the huge amount of social network users, representative samples can be easily obtained from any country; (ii) given the facility for accessing user profiles in social networks, values for more specialized groups can be obtained by segmenting respondents using features like age, sex, religion, or education level; and (iii) responses are constantly generated, which provides fresh values for practical purposes. Figure 12.1 shows new values obtained from a sample of 600 Mexican Facebook users segmented by age and sex.

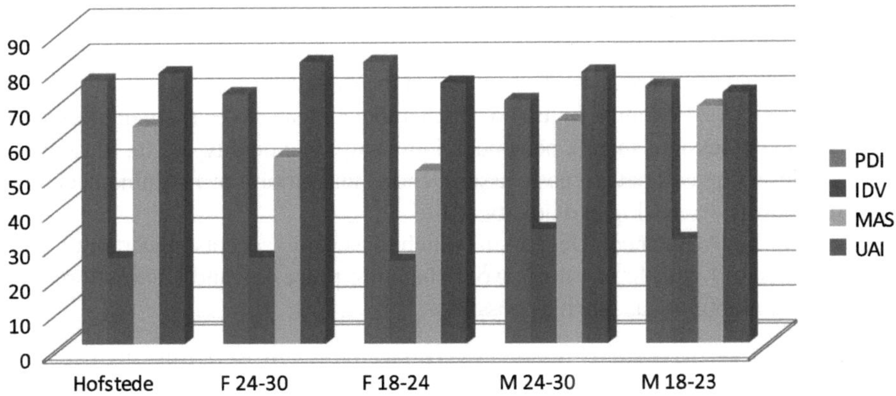

Fig. 12.1 Mexican cultural values segmented by age and sex

12.2.2 Hofstede's Model in Practice

The model of Hofstede has been used mainly for two purposes: (i) for technological developments, where the model has driven the design of some applications; and (ii) as a theoretical framework for analyzing common conflicts when individuals from different cultures interact. The platform proposed is a combination of these two points of view: the model of Hofstede is the foundation for creating cultural profiles for users and theory shows some conflicts which the platform intends to manage.

From the technological point of view, there are some developments influenced by the model. Nazir et al. (2000) developed an agent model for managing intercultural interactions which combines culture, personality and emotions. Personality and culture influence emotional needs of the agent which maintains a long term memory with rigid cultural rules and a short term memory with information about behaviors expected from other agents. When such expectations do not match with events in the environment the agent updates his short term memory in order to adapt.

Rehm et al. (2007) propose a model for deriving the cultural background of a user based on his behaviors and adapt the culture of an embodied conversational agent. Basically, the system classifies the culture of the user into one of eight stereotypes. This is done by capturing movements of the user, analyzing observations, and matching them to one of the cultural stereotypes. The adopted culture affects aspects like spatial extension, speed, and power of movements. Hofstede dimensions are used as an entry point in a Bayesian network while deriving the cultural background of the user, and when assigning behaviors to the agent according to the adopted culture.

Mascarenhas et al. (2009) developed a model of culture based on individualism and power distance values. In this model, goals and emotions of agents are

Table 12.1 Examples of cultural conflicts

1	*Individualism*	*Collectivism*
1.1	Decisions are taken individually	Decisions are taken in group
1.2	Individual points of view must be expressed	Confrontation must be avoided
1.3	Own goals are primordial	Goals of the group are primordial
2	*Low power distance*	*High power distance*
2.1	Subordinates have liberty in their work	Subordinates are told what to do
2.2	Anyone can initiate interactions if necessary	Interactions are supposed to be initiated by superiors only
2.3	Subordinates are consulted	Subordinates are notified
3	*Masculinity*	*Feminity*
3.1	Majority is sufficient	Consensus is desirable
3.2	Success depends on results	Success depends on relationships and teamwork
3.3	Competitiveness is valued	Modesty and cooperation are valued
4	*Low uncertainty avoidance*	*High uncertainty avoidance*
4.1	General aversion against rules	General necessity of rules
4.2	Hard work only if necessary	Hard work is a constant
4.3	Calendars are flexible	Punctuality is essential

influenced by its cultural values. Individualism and power distance values of the agent affect utility of goals according to Hofstede's descriptions about how different values on such dimensions modify behaviors. Emotions are influenced in the same way, by means of an equation based on theory about the influences of individualism and power distance.

From the theoretical perspective, there are many examples of cultural conflicts which are analyzed using Hofstede's dimensions. Commonly, such conflicts are related to a dimension and to a context of interaction. Table 12.1 shows some common situations which led to conflicts in intercultural collaboration; these are the kind of conflicts the platform is intended to manage.

12.3 Proposal Overview

Our objective is to provide a multi-agent system for supporting intercultural collaboration and reducing occurrences and impacts of cultural misunderstandings. Our approach bases on cognitive agents with knowledge about the culture of users, which analyze interactions of users searching for the occurrence of potential cultural conflicts. Interactions are formalized according to the language-action perspective (LAP) (Winograd and Flores 1987), whose main foundation is that people act through language. Using LAP, interactions are modeled as conversations which evolve as users perform actions until reaching a final state. The set of possible actions is extracted from speech act theory and the taxonomy proposed by Searle (1975).

12.3.1 Multi-agent System Architecture

The proposed MAS is based on OMAS (Barthès 2011), a platform for developing cognitive agents. Some of the capabilities of OMAS agents which make them suitable for our purposes are:

- Autonomous execution and proactive behavior: OMAS agents execute autonomously in the background and respond to stimulus by means of their skills. Skills are predefined behaviors which can be executed by an agent when it receives a request from other agent or from another element in the environment. They can also display goal-oriented behaviors; an agent possesses a predefined set of goals, and given a scenario they can construct a plan for achieving them, which can involve activities from other agents. Furthermore, an agent can build goals dynamically.
- Structuring and Storing Knowledge: OMAS agents structure their knowledge by means of ontologies, which are formalized with the MOSS knowledge representation language. They include definitions of different ontologies related to their built-in functionalities, and also can be provided with other ontologies required for the specific functionality of each agent.
- Perception of the Environment: OMAS agents can perceive contextual information in order to update their knowledge and adapt their behaviors accordingly. Such perception is represented as incoming messages. As an example, users can interact with personal assistants through a vocal interface, and user utterances are delivered as messages to personal assistants.
- Reasoning Capabilities: cognitive agents possess inference mechanisms which allow them to produce new knowledge by reasoning over their knowledge base. OMAS agents have a persistent store for keeping their knowledge bases and can make inferences over them. In addition, the MOSS formalism allows performing specific queries which can be answered by the same agent or sent to other agents.
- Learning Capabilities: OMAS agents can implement skills for learning from experience. With such skills, OMAS agents can evaluate the results of certain actions, and they can adapt their behaviors in order to improve performance.
- Collaboration: OMAS agents can work together in order to reach common goals. They can send and receive messages, and in this way they share information and make requests for executing skills. In addition, OMAS agents are aware about the existence and the capabilities of other agents, and they can generate plans which involve the execution of skills from several agents.

12.4 Formalizing Culture

This section describes cultural profiles, the means proposed for representing culture of users. Cultural profiles are based on the model proposed by Hofstede, however we decided to represent cultural dimensions using linguistic variables rather than numeric variables. Zadeh (1975) defines linguistic variables as *a formalism for dealing with complex and ill-defined systems, whose behavior is strongly influenced by human judgment, perception, or emotions*. This definition fits appropriately with the concept of culture: there is no a rigid definition about culture and how to measure it, and the perception about cultural preferences is completely dependent of judgment.

Usage of linguistic variables allows profiling someone with values like *very individualistic* or *more or less risky*, which is more natural than numeric values. Linguistic variables are also convenient when defining cultural differences, using them it is possible to approximate differences in cultural dimensions using fuzzy relations like *greater than* or *approximately equal than*. Finally, usage of linguistic variables allows using fuzzy logic and approximate reasoning for comparing cultural profiles and making inferences from their values.

12.4.1 Linguistic Variables

A commonly used example for illustrating linguistic variables is *age*. While a typical variable for ages holds integer values, a linguistic variable holds text values like *young, quite young, not very young and not very old*, etc.

A linguistic variable is formally defined as a quintuple $L = (\vartheta, T(\vartheta), U, G, M)$, where:

- ϑ is the name of the variable, e.g. *age*;
- $T(\vartheta)$ or the term set, is the collection of possible linguistic values;
- U is the universe of discourse, i.e. the base variable from which linguistic values are defined. For the example of ages, $U = \{0, 1, 2, \ldots, 100\}$;
- G is a syntactic rule generating terms in $T(\vartheta)$, usually given by a context-free grammar;
- M is a semantic rule, which represents the meaning of linguistic values in $T(\vartheta)$. Semantics of a linguistic variable is given by a set of compatibility functions, one for each possible linguistic value X. A compatibility function $c_X : U \to [0, 1]$ denotes the compatibility of each $u \in U$ with the linguistic value X. For example, some evaluations for the compatibility function associated with the linguistic value *young* would be: $c_{young}(10) = 1, c_{young}(27) = 0.7, c_{young}(35) = 0.2, c_{young}(90) = 0$.

Elements in the term set are either primary terms, like *young, old, middle-aged*, or composite linguistic values composed by primary terms, hedges (*very, more or less, quite*, etc), and connectives (*and, or, not*). In general, system designers heuristically define compatibility functions for primary terms, while values of hedges and connectives are computed with nonlinear operators.

Compatibility functions for linguistic values containing hedges are often approximated using exponential functions (Zadeh 1972). For example the hedge *very*, which has an intensifying effect, is commonly approximated by the square function, while *more or less*, which has the opposite effect, is commonly approximated by the square root function.

$$c_{very\,X}(u) = c_X(u)^2,$$
$$c_{more\ or\ less\,X}(u) = c_X(u)^{1/2}$$

In the case of connectives, compatibility functions are approximated with functions that must satisfy a set of specific axioms (Lee 2005). The most common functions are:

$$c_{not\,X}(u) = 1 - c_X(u),$$
$$c_{X\ and\ Y}(u) = min(c_X(u), c_Y(u)),$$
$$c_{X\ or\ Y}(u) = max(c_X(u), c_Y(u))$$

Note that presented functions are general and apply for any linguistic variable. On the other hand, compatibility functions for primary terms are specific for each linguistic variable and in most cases they are defined and tuned heuristically.

12.4.2 Cultural Profiles

Cultural dimensions measure dichotomies, i.e. the extent in which people prefer one of two possible states. Bounding values represent completely opposite stereotypes, and center values represent balance in preferences.

We define a cultural profile as a set of linguistic variables, each of them modeling a cultural dimension. Such linguistic variables are defined over the universe [0, 100], the domain proposed by Hofstede.

For each linguistic variable v_i in the profile, three primary terms are defined: p_{i1}, p_{i2}, and *neutral*. The terms p_{i1}, and p_{i2} represent both stereotypes, and they are represented as two sigmoid shapes biased to 0 and 100 respectively and symmetric with respect to $x = 50$. *Neutral*, is present in all cultural dimensions and allows profiling individuals whose behaviors are not clearly biased towards a stereotype. The shape of the compatibility function for *neutral* follows a distribution around $x = 50$. The shape of compatibility functions for stereotypes is based on item response theory (Baker and Kim 2004), which is used for analyzing data obtained from measurements of things like abilities, attitudes, and personality traits. In particular, compatibility is characterized using the logistic function, given by:

$$f(u) = 1/(1 + e^{-a(x-b)}), \text{ where}$$

- b is the position of the center of the curve, and
- a determines the smoothness of the curve, which is proportional to its maximum slope, i.e. to the slope of the curve in b

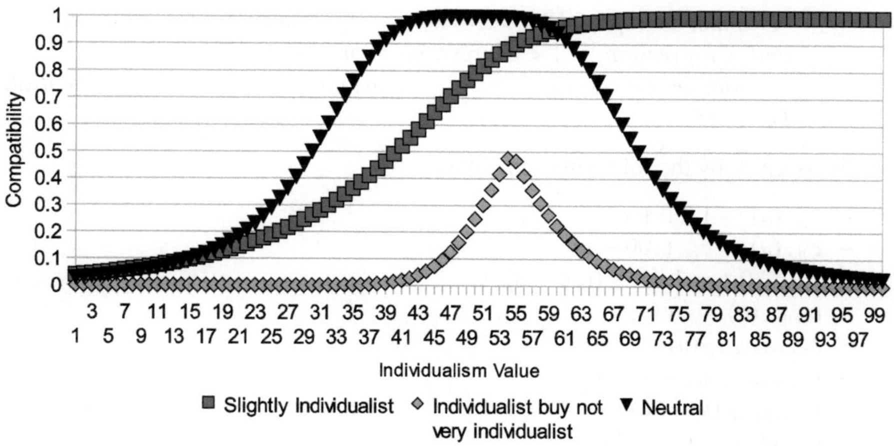

Fig. 12.2 Compatibility functions for collectivism/individualism

Finally, the compatibility function for the linguistic value *neutral* is given by:

$$c_{neutral}(u) = 1/[1 + ((50 - u)/d)^4],$$

where d is proportional to the width of the curve. The set of hedges proposed is $H = \{somewhat, moreorless, very, extremely\}$. Figure 12.2 shows compatibility functions for some linguistic values of the *individualism/collectivism* dimension.

Formally, for a given application influenced by n cultural dimensions, culture is modeled as a set:

$$C = \{v_1, v_2, v_3, \ldots, v_n\}$$

Where each v_i is a linguistic variable representing a cultural dimension, defined as:

$$v_i = (\vartheta_i, T(\vartheta_i), U, G_i, M_i), \text{where:}$$

- ϑ_i is the name of the cultural dimension;
- $T(\vartheta_i)$ is the term set, generated by G_i;
- $U = \{0, 1, 2, \ldots, 100\}$;
- $G_i = (W_i, \sum_i, R_i, \sigma_i)$ is a context-free grammar with initial symbol σ_i, where
  - $W_i = p_{i1}, p_{i2}, neutral \cup H$
  - $\sum_i = \{\sigma_i, < expression_i >, < simple\_expression_i >,$
    $< connective\_expression_i >\}$
  - R_i is the following set of productions:
    - $\sigma_i ::=< expression_i >,$
    - $< expression_i >::=< simple\_expression_i > | < connective\_expression_i >,$

- $<$ *connectiveexpression*$_i$ $>$::= *not* $<$ *expression*$_i$ $>$ | $<$ *expression*$_i$ $>$ *and* $<$ *expression*$_i$ $>$ | $<$ *expression*$_i$ $>$ *or* $<$ *expression*$_i$ $>$,
- $<$ *simple_expression*$_i$ $>$::= $p_{i1}|h_j\, p_{i1}|p_{i2}|h_j\, p_{i2}|neutral|h_j neutral$; $h_j \in H$

- M_i is given by the following functions:
  - $c_{p_{i1}}(u) = 1/(1 + e^{-a_i(u-b_i)})$
  - $c_{p_{i2}}(u) = c_{p_{i1}}(100 - u)$
  - $c_{neutral}(u) = 1/[1 + ((50 - u)/d_i)^4]$
  - $c_{very\,X}(u) = c_X(u)^2$
  - $c_{more\ or\ less\ X}(u) = c_X(u)^{1/2}$
  - $c_{somewhat X}(u) = c_X(u)^{1/3}$
  - $c_{extremely X}(u) = c_X(u)^3$
  - $c_{not\ X}(u) = 1 - c_X(u)$
  - $c_{X\ and\ Y}(u) = min(c_X(u), c_Y(u))$
  - $c_{X\ or\ Y}(u) = max(c_X(u), c_Y(u))$

According to this definition, defining a cultural dimension only requires assigning values for parameters a_i, b_i, and d_i, which determine the shape of the compatibility functions.

12.4.3 Values for Culture

The main advantage of modeling cultural profiles as cultural dimensions is that they can be measured for any given user. However, assigning values to cultural profiles is not straightforward and requires a deep understanding of the dimensions in question.

Theories proposing cultural dimensions associate each extreme of dichotomies with typical behaviors. Such behaviors are analyzed in order to develop feasible approaches for measuring cultural dimensions. Hofstede publishes in his website values for his cultural dimensions for 56 different countries and regions (Hofstede, www.geert-hofstede.com). Such values are the result of statistical analysis of tests applied during his research. Another example is the approach adopted by Hall in proxemics theory, providing values for physical distances among humans based on an analysis of perception (sight, hearing, smell, touch, thermoception), categories of interactions, and observation of behaviors of different cultures. Linguistic variables give the opportunity of profiling users heuristically based on the available information. Internally such values are managed as numerical values, for which they are processed by computing their membership functions and applying a defuzzification method.

12.5 Interactions and Conflicts

This section presents the approach adopted for formalizing interactions among users and defining cultural conflicts. Such approach is an extension of the language action perspective (LAP) (Winograd and Flores 1987). The main idea behind LAP is that people act trough language. As consequence of this idea, the LAP is closely related with the speech act theory (Austin 1975). In LAP, interactions are treated as conversations among users, where each turn in the conversation is characterized by an illocutionary act. Each conversation is represented as a state transition network, where user actions trigger transitions until reaching a final state.

We propose extending the model by allowing the specification of cultural conflicts in each state of the conversation. Conflicts are composed by a condition and a reaction. Conditions combine fuzzy and Boolean logic; they express differences of cultural profiles, and comparisons of contextual information. When a new stage is reached in a conversation, the corresponding conditions are evaluated, and if it is the case, the reaction is triggered.

The LAP approach presents some convenient features for our purposes: is intuitive, so common users can define new patterns of interaction; diversity of illocutionary acts allows modeling complex and varied patterns of interaction; and usage of state transition networks allows defining conflicts in any point of an interaction.

12.5.1 The Language-Action Perspective

The language action perspective is an approach for supporting communication and coordinated action among groups of people. Unlike traditional approaches which give special attention to information processing and data transfer, LAP considers language as the main dimension of cooperative work. LAP bases on the illocutionary force of utterances, i.e. in the action that is actually performed each time we participate in a conversation. For example, by pronouncing an utterance we can promise, request, apologize, declare, make an offer, make a counteroffer, ask, etc.

Under the LAP approach, common patterns of interaction are modeled as conversations for action (CFA). CFAs represent the different paths, in terms of illocutionary acts, that can be followed in a specific kind of interaction. A simple example presented by Winograd (Winograd and Flores 1987) is the CFA where user A makes a request to user B. Figure 12.3, obtained from Winograd and Flores (1987), shows the state transition diagram of such interaction. This pattern of interaction starts when user A makes a request; then user B can accept, decline or make a counter-offer; each of this options provides new possibilities of action for A, then B has new possibilities, and so on until reaching a final state.

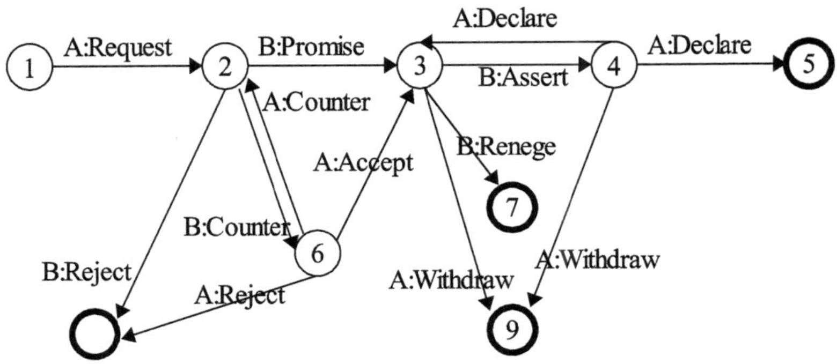

Fig. 12.3 A CFA for a request-promise-response interaction

12.5.2 Defining Cultural Conflicts

We propose extending CFA for defining conflicts prone to occur in interactions. Conflicts are defined by adding *if-then* rules within states of CFAs. Conditions of rules allow taking in account diverse kinds of information like: differences in cultural profiles, personal information about users, and metadata about messages.

Each condition is associated with a reaction. Reactions are tasks performed by cultural broker agents and can be supported by additional staff agents. Therefore they may vary from quite simple behaviors, like showing messages to users, to very complex ones, like executing fuzzy algorithms and updating cultural profiles.

The elements identified for defining cultural conflicts can be divided in four main categories:

- Personal Information. In addition to cultural values, users can be profiled with information like their age, sex, academic degree, position in the organization, etc. The fields to be added depend on the application and comprise all the information that is required for defining the corresponding conflicts. For example, in conflicts related to power distance, it is useful to have information about the position of users in the organization; in conflicts related to *masculinity/femininity*, it is useful to know the sex of users.
- Timing of Interactions. CFA allows modeling interactions as conversations where users interact in turns. Cultural preferences of a user may be reflected in the amount of time he takes for responding when his turn comes. For example, the delay in responses may be useful for defining conflicts related to *individualism*, and the CBA could tune such value accordingly.
- Metadata of Messages. In the CFA model, actions are captured only by their illocutionary forces. However for some conflicts it would be helpful to describe actions with further information. Such kind of information can refer to the

content, or to the nature of the message. Characterizing the content might be done by adding related concepts from an ontology. Characterizing the nature of the message can be done by adding several kinds of information, like the level of priority of the message, the maximum deadline in which a response is expected, or the strength (Searle 1975) of the illocutionary force. For example, for defining conflicts related to the *Long/Short term orientation* dimension, it would be useful to know the time deadline in which a response is expected; or for defining conflicts related to uncertainty avoidance, information about the priority of messages may be required.

- Cultural Differences. For expressing cultural conflicts, it is fundamental to allow the expression of cultural differences. The framework provides two ways for doing so: using assignations on linguistic values directly or using fuzzy relations among them.

The most natural way is by using linguistic values directly, by operating values between two users using Boolean operators. For example, a conflict respecting with *individualism/collectivism* could be defined using *if A is very individualist and B is very collectivist* as part of the condition.

The second possibility is to use fuzzy relations for expressing relative differences in cultural profiles. We use a linguistic variable called *comparison*, which allows defining conflicts using conditions like *if collectivism of B is much greater than collectivism of A*.

The primary terms defined for the variable *comparison* are *higher, lower, and equal*. The membership function of each of these values is a fuzzy set defined over the domain $[0, 100] \times [0, 100]$, i.e. the product of two domains of cultural dimensions (two domains, because *higher, lower*, and *equal* are binary operators).

As mentioned, primary terms may be combined using linguistic hedges; for *greater* and *lower*, allowed hedges are *slightly, more or less, much, extremely*; for *equal*, allowed hedges are *hardly, more or less, almost*.

The membership functions for *greater* and *lower* are based on a threshold function:

$$c_{lower}(x, y) = min(1, [(x - y)/a]^2), x > y$$

$$c_{greater}(x, y) = 1 - min(1, [(x - y)/a]^2), x < y$$

where the parameter a determines the steepness of the threshold.

The membership function for *equal* is based on a function for defining ranges:

$$c_{equal}(x, y) = 1/[1 + ((y - x)/b)^4]$$

where the parameter b determines the shape of the curve.

Figure 12.4 shows the compatibility functions for *greater*, *lower* and *equal*, and illustrates the role of the parameters.

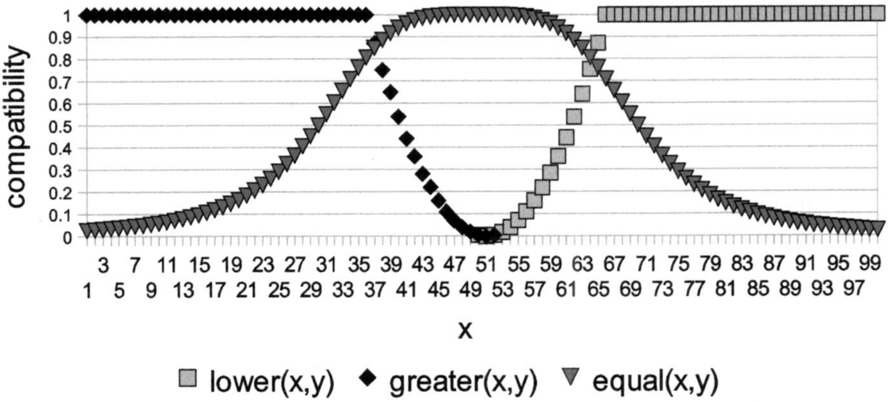

Fig. 12.4 Primary terms for the comparison fuzzy relation

12.5.3 Examples of Cultural Conflicts

Next, we present some illustrative examples of cultural conflict definitions. The examples presented are taken from Table 12.1 and adapted to the CFA showed in Fig. 12.3.

12.5.3.1 Expressing Contradictory Points of View

According to conflict 1.2 (Table 12.1), individualist users express points of view, even if they are contradictory. On the other hand, it is not common for collectivist users to express contradictory points of view in order to avoid confrontations. In the context of the CFA in Fig. 12.3 (supposing B is collectivist), user B could response with a promise, even one that he knows that is hard to fulfill. The consequences of such conflict are considerable when a request has high priority. Such conflict is defined in the knowledge base of the CBA supporting user B, specifically within state 2 of the definition of the CFA. One possible way of defining such conflict is:

if B is highly collectivist and request_priority = "high"

In such case the agent could react, for example by advising the user for responding honestly (i.e. with a counteroffer or a withdraw). Note that the definition of this conflict requires that the user A provides the priority of the request (metadata in messages).

12.5.3.2 Response Time

According to conflict 1.3 (Table 12.1), importance given to own and group goals differs with different degrees of individualism/collectivism. According to this, collectivist users are likely to respond and expect responses more quickly than individualist users. In the context of the CFA in Fig. 12.3 (supposing B is individualist and A is collectivist), if user B has taken more than a certain time limit for making a promise, it is likely that he will take a long time also for accomplishing the promise. Such conflict is defined in the knowledge base of the CBA supporting user A, specifically within state 3 of the definition of the CFA. One possible definition of such conflict is:

if B is individualist and A is collectivist and
timestamp(transition2)-timestamp(transition1)>limit

In such case the agent could react, for example by advising user A for withdrawing 3. Note that the definition of this conflict requires the usage of transition timestamps.

12.5.3.3 Starting Conversations

According to conflict 2.2 (Table 12.1), in collectivist cultures interactions are supposed to be initiated by superiors only. In the context of the CFA in Fig. 12.3 (supposing power distance of A is much lower than power distance of B), if user A starts an interaction and B is a superior, B may get upset. Such conflict is defined in the knowledge base of the CBA supporting user A, specifically within state 1 of the definition of the CBA. One possible definition of such conflict is

if power_distance of A is much lower than power_distance of B and
superior(B,A)

In such case the agent could react, for example by advising user A for not sending the request at least it is really important. Note that this definition requires that the CBA maintains knowledge about the position of users in the organization (personal information).

12.6 Evaluating Conflicts

As described in the previous section, the condition of a conflict contains two main components: a fuzzy component for expressing cultural differences (in terms of linguistic values), and a non-fuzzy component for describing required additional contextual constraints. The evaluation of contextual constraints is done using Boolean logic over traditional primitive data types. This section focuses on the evaluation of cultural differences.

12.6.1 Defuzzification

Values for cultural dimensions are expressed as linguistic variables either if they are heuristically assigned to the cultural profile of a user or if they are expressed in conditions. Such linguistic values must be converted to numbers in order to allow computations. This process is called defuzzification and there exist several methods for performing it. The most widely used is by obtaining the *centroid* of the compatibility function:

$$z_X = \left(\sum_{u \in U} c_X(u) \cdot u\right) / \left(\sum_{u \in U} c_X(u)\right)$$

For example, the defuzzification process of the linguistic value *Individualist but not too individualist* with the parameters $a = 0.25, b = 70, c = 10$, is the following:

$$c_{ind}(u) = 1/(1 + e^{-0.25(u-70)})$$

$$c_{very\ ind}(u) = c_{ind}(u)^2$$

$$c_{not\ very\ indi}(u) = 1 - c_{very\ ind}(u)$$

$$c_{ind\ but\ not\ very\ ind}(u) = min(c_{ind}(u), c_{very\ ind}(u))$$

$$z_{ind\ but\ not\ very\ ind} = \sum_{u=0}^{100} c_{ind\ but\ not\ very\ ind}(u) \cdot u / \sum_{u=0}^{100} c_{ind\ but\ not\ very\ ind}(u)$$

$$z_{ind\ but\ not\ very\ ind} = 74.52$$

12.6.2 Evaluating Linguistic Values

This section presents the method for computing the probability of occurrence of a conflict defined with direct assignations of linguistic values. This section is exemplified by evaluating the condition *if B is individualist and A is collectivist*, contained in the conflict about response times presented in Sect. 12.5.3.

The first step is to deffuzify the value of all the linguistic values present in the condition. With the same parameters used above we have:

$$z_{ruleA} = z_{ind} = 15.61$$

$$z_{ruleB} = z_{col} = 84.39$$

Then the actual cultural values of the involved users are defuzzyfied. Suppose that user A is profiled as *Individualist but not very individualist* and user B is profiled as *extremely individualist*.

$$z_{userA} = z_{indbutnotveryind} = 74.52$$

$$z_{userB} = z_{extremelyind} = 87.59$$

Then the average difference of profiles between the condition and the actual cultural profiles of involved users is computed.

$$avg = (|z_{ruleA} - z_{userA}| + |z_{ruleB} - z_{userB}| + \ldots + |z_{rulen} - z_{usern}|)/n$$

For the values given in the example:

$$avg = (58.91 + 3.2)/2 = 31.055$$

Small values for *avg* mean a close correspondence with the rule. Therefore, the smaller the value of *avg*, the bigger the fulfillment of the condition, and therefore the bigger the probability of occurrence of the conflict.

12.6.3 Evaluating Relative Differences

The process for computing the probability of occurrence of a conflict defined using relative differences is similar to the own presented in Sect. 12.6.2. This section is exemplified by evaluating the condition *if power_distance of A is much lower than power_distance of B*, contained in the conflict about starting conversations in Sect. 12.5.3.

The first step is to deffuzify the actual cultural values of the two involved users: z_{userA} and z_{userB}. Suppose that user A is profiled as *very equitative* and user B is profiled as *more or less neutral.*

$$z_{userA} = 13.44$$

$$z_{userB} = 50$$

Then the compatibility function of the involved fuzzy relation is obtained. For the example such relation is *much lower*.

$$c_{lower}(x, y) = min(1, [(x - y)/a]^2); x < y c_{much\ lower}(x, y) = c_{lower}(x, y)^2; x < y$$

Then the function is evaluated for $x = z_{userA}$ and $y = z_{userB}$, assuming $a = 40$:

$$c_{much\ lower}(13.44, 50) = (min(1, [(13.44 - 50)/40]^2))^2 = 0.6979$$

The bigger the value obtained, the bigger the degree of fullfilment of the condition, and therefore the bigger the probability of occurrence of the conflict.

12.7 Conclusions

This chapter presents a framework for managing conflicts in multicultural interactions. The proposal presented represents a first attempt for developing a system intended at reducing the impacts of cultural differences in multicultural collaboration; it is based on constructing quantifiable cultural profiles which are based on strong theoretical basis.

Cultural conflicts are defined based on formalized interactions represented as patterns of actions. The framework provides a flexible and application independent means for defining conflicts in terms of differences in cultural profiles and other kinds of contextual information. Definition and evaluation of cultural differences are based on theories of linguistic variables, fuzzy logic, and approximate reasoning, which improves intuitiveness and simplicity of usage.

The platform consists in a multi-agent system composed of cognitive agents in charge of supporting interactions, maintaining definitions of cultural conflicts, and reacting accordingly as users interact. The framework has been designed based on the theory of Hofstede, respecting to both, its model of culture and the conflicts analyzed under his model.

References

Austin, J.L.: How to Do Things With Words. Harvard University Press, Cambridge (1975)

Baker, F., Kim, S.H.: Item Response Theory: Parameter Estimation Techniques, CRC Press (2004)

Barthès, J.P.: OMAS - A Flexible Multi-Agent Environment for CSCWD. Future Generation Computer Systems Vol 27 Issue 1, pp. 78–87 (2011).

Bryman, A.: Quantity and Quality in Social Research, Routledge, London, 1988.

De Troyer, O., Mushtaha, A., Stengers, H., Baetens, M., Boers, F., Casteleyn, S., Plessers, P.: On Cultural Differences in Local Web Interfaces. Journal of Web Engineering Vol. 5 Nr. 3, pp.246–264 (2006)

Dutsbar, S., Hofstede, G.J.: Videoconferencing across cultures: A conceptual framework for floor control issue. Journal of Information Technology Vol. 14, pp.161–169 (1999)

Hall, E.T.: The Dance of Life: The Other Dimension of Time. Anchor Books, New York (1989)

Hall, E.T.: The Hidden Dimension. Anchor Books, New York (1990)

Hall, E.T.: The Silent Language. Anchor Books, New York (1999)

Hill, C., Jones, G.R.: Strategic Management an Integrated Approach. South-Western College Pub, Boston (2006)

Hofstede, G., Bond, M.H.: The Confucius Connection from Cultural Roots to Economic Growth. Organizational Dynamics Vol 16 Issue 4, pp.5–21 (1988)

Hofstede, G.: Cultures and Organizations: Software of the Mind. McGraw Hill Inc, New York (1996)

Geert Hofstede, www.geert-hofstede.com

Kamel, N., Davison, R.: Applying CSCW technology to overcome traditional barriers in group interactions. Information Management Vol. 34, no. 4, pp. 209–219 (1998)

Lee, K.H.: First Course on Fuzzy Theory and Applications, Springer(2005)

Mascarenhas, S., Dias, J., Prada, R., Paiva, A.: One for All or One for One? The Influence of Cultural Dimensions in Virtual Agents' Behaviour. In: Proceedings of the 9th International Conference on Intelligent Virtual Agents (IVA 09), pp. 272–286, (2009).

McSweeney, B.: Hofstede's model of national cultural differences and their consequences: a triumph of faith, a failure of analysis. Human Relations Journal, pp. 89–118 (2002).
Nazir, A., Enz, S., Lim, M.Y., Aylett, R., Cawsey, A.: Culture-Personality Based Affective Model, AI and Society Vol. 13 Nr. 3, pp. 281–293, (2000)
Llobera, J., Spanlang, B., Ruffini, G., Slater, M.: Proxemics with Multiple Dynamic Characters in an Immersive Virtual Environment. ACM Transactions on Applied Perception Volume 8 Issue 1, pp.38–50 (2010)
Paez, D., Fernandez, I., Ubillos, S., Zubieta, E.: Social Psichology, Culture and Education. Prentice Hall, Madrid (2003)
Rehm, M., Bee, N., Endrass, B., Wissner, M., André, E.: Too Close to Comfort?: adapting to the user's cultural background. In: Proceedings of the international workshop on Human-centered multimedia, pp. 85–94, Germany (2007).
Searle, J.: A Taxonomy of Illocutionary Acts. In: Minnesota studies in the philosophy of language, pp.334–369. University of Minnesota Press, Mineapolis (1975)
Schwartz, S.H.: A Proposal for Measuring Value Orientations Across Nations. In:The Questionnaire Development Package of the European Social Survey (2003)
Trompenaars, F.: Riding the Waves of Culture: Understanding Diversity in Global Business. McGraw Hill Inc (1998)
Winograd, T., Flores, F.: Understanding Computersand Cognition: A New Foundation for Design. Addison Wesley, Massachussets (1987)
Zadeh L.A.: A Fuzzy-Set-Theoretic Interpretation of Linguistic Hedges. Journal of Cybernetics Vol. 2 Nr. 3, pp. 4–34, (1972)
Zadeh, L.A.: The Concept of Linguistic Variable and its Application to Approximate Reasoing. Information Sciences, 8, pp. 199–249 (1975)

Chapter 13
The Culture Driven Game Design Method: Adapting Serious Games to the Players' Culture

C.J. Meershoek, R. Kortmann, S.A. Meijer, E. Subrahmanian, and A. Verbraeck

13.1 Introduction

In the process of interaction, the players of serious games will always bring their own culture into the game (Consalvo 2009; Fine 1983). Practice showed that if the game is not aligned with the culture of the players, this can result in conflicting behaviour that hampers the players to reach the objectives of the game. To solve this issue the design of the game architecture needs to be adjusted. A method was developed and tested in a collaboration project of the Delft University of Technology in the Netherlands and the Centre for Study of Science, Technology and Policy from Bangalore, India. This chapter proposes the new Culture Driven Game Design Method which supports serious game designers in adapting their game design to the culture of the players. The Culture Driven Game Design Method provides a tool to assess and represent the culture of the targeted players as well as a method to process this assessment and avoid conflicts between the culture of the players and the architecture of the game.

Let us demonstrate the effect of culture in serious games by providing two examples from the field. The first example comes from Germany where months of careful

C.J. Meershoek (✉)
Be Involved, The Hague, The Netherlands
e-mail: cjmeershoek@beinvolved.nl

Delft University of Technology, Delft, The Netherlands

R. Kortmann • S.A. Meijer • A. Verbraeck
Delft University of Technology, Delft, The Netherlands

E. Subrahmanian
Delft University of Technology, Delft, The Netherlands

Center for Study of Science, Technology, and Policy, Bangalore, India

Carnegie Mellon University, Pittsburgh, USA

V. Dignum and F. Dignum (eds.), *Perspectives on Culture and Agent-based Simulations*, Studies in the Philosophy of Sociality 3, DOI 10.1007/978-3-319-01952-9_13,

and expensive preparation were put in the supply chain game that was set up for a full afternoon of gameplay by a company department. Challenges in supply chain management arise due to a lack of information availability throughout the chain.

The game was made to let the participants experience the consequences of this information scarcity. Despite the extensive preparation, the game was finished in less than 5 min after the department boss summoned each player to provide all the information available in the game to him (Meijer et al. 2006). As a result of this action, the objectives of the game were not met.

Another example involves a trading game designed at an American university. When this game was played with American students it did not last long. The opportunity to let other players go bankrupt was immediately interpreted as the objective in the game. When the game was played with exchange students from Taiwan it took hours and hours before the game was eventually aborted. At the time the game was ended, none of the players had gone bankrupt. The opportunity to let other players go bankrupt was not interpreted as the objective in the game by the Taiwanese students. If any player was low on cash, he was helped by other players so that bankruptcy was avoided. The teachers were stunned by this result of the game (Mayer, personal communication).

The common factor in these examples is that the players were able to play the game within the set of given rules but it still resulted in highly unexpected behaviour. This implies that the group of players had a shared basis of unwritten rules that structured their behaviour during the game that was unknown by the designer and facilitator. This shared basis of unwritten rules can be dubbed the culture of that specific group of players (Caluwé et al. 2008).

This culture-related behaviour changed the games in such a way that it is unlikely that the objectives of the game have been met in these sessions. There is no need in spending resources on a serious game if the objectives of the game cannot be reached. These examples in fact emphasize the conclusion drawn in the work of Hofstede (2008); cultural aspects of serious games are of paramount importance to the acceptance and successful learning outcomes of simulation gaming sessions.

These conclusions affect serious gaming as a tool for complex multi-actor problems. Serious gaming is an important tool in creating, explaining, building, deploying and evaluating solutions for complex multi-actor problems (Abt 1970; Duke 1974; Duke and Geurts 2004; Klabbers 2008; Mayer 2009; Mayer and Veeneman 2002). Serious gaming provides the opportunity to interact with complex models and experience (r)evolutionary changes (Mayer 2008). By doing this in a game, solutions can be implemented and tested without damaging the real world (Abt 1970). This is a great benefit in a context of complex multi-actor problems (Mayer 2008).

In order to avoid scenarios as sketched earlier, serious games need to be adapted to the culture of the targeted players. It is possible to adapt serious games to the culture of the targeted players by playtesting with these players (Fullerton 2008). Playtesting is the iterative process in which the game is designed, tested and evaluated, each time improving the game, until the player experience meets your criteria (Fullerton 2008). However, practice shows that playtesting with the targeted players is not always possible or desirable from the game designers point of view for two reasons. First, the development of new games is very costly and time-consuming (Duke and Geurts 2004), it may therefore not be possible to organize and facilitate such a test play session with all the targeted players, especially if that target group consists of busy, expensive, high-ranking officials. Second, playtesting with the targeted players is undesirable from the point of view of the designers of the game because they want to make a good first impression. (e.g. for funding reasons) So adapting serious games to the culture of the players through direct playtesting is not always an option.

In search of alternative ways to adapt games to the players' culture existing serious game design methods, such as (Crawford 1984; Duke and Geurts 2004; Fullerton 2008; Kortmann and Harteveld 2009), were analyzed. None of these methods provide an alternative to the playtest method. It is therefore that this chapter proposes a new method that is able to adapt serious games to the culture of the players without playtesting it with the targeted players; the Culture Driven Game Design Method.

13.2 Culture in Serious Games

Before describing the Culture Driven Game Design Method, this section provides background and demarcates the problem by means of the theoretical basis of this research. The theoretical basis consists of two interrelated frameworks of (Meijer 2009) and (Williamson 2000). The first framework describes the inputs and outputs of a serious game session. Using this framework the relation between a serious game session and the culture of the players can be explained. The second framework describes this culture and integrates it with the different environments in which complex multi actor problems are dealt with. After the description of the two frameworks, this theoretical basis was used to structure the demarcation of this research.

The first framework of the theoretical basis of this research is a model of (Meijer 2009) which provides an overview of all the inputs and outputs of a gaming session. This model is briefly described here, whereas a more extensive explanation can be found in (Meijer 2009).

In order to play a session with *participants*, a *design* and a *configuration* are needed. The outputs of the session are *quantitative and qualitative data* together with the *experience* the participants gained during the session. Part of what the participants bring to the game is their personality, their relational history and their culture[1] (Meijer 2009). This research is focussed on the influence of culture in serious games.

---

[1]It should be noted that this statement, and thereby the theoretical basis of this research, conflicts with the theory of the magic circle. The magic circle is a widely used theoretical concept introduced by (Huizinga 1955) which claims that the world in which a game is played is completely isolated from the real world (Harvey 2006; Paras and Bizzocchi 2005; Salen and Zimmerman 2004). In this research (Consalvo 2009; Fine 1983) are followed who both concluded that the real world will always intrude into the gameplay.

It should be noted that the behaviour of participants cannot be explained by culture alone, factors as relational history and personality also influence behaviour. As a consequence culture cannot be observed directly, which is why in this research a group test is used as a proxy for the collective parts. It is acknowledged that such a group test will not reveal whether the collective parts stem from personality or culture. However, a group test as such is considered the best proxy for culture.[2]

Culture exists at national, regional and corporate levels (Watson et al. 1994). Culture is associated with beliefs, norms, mores, myths, value systems and structural elements of a specified group of people (Nath 1988). This implicates that not contesting your superior as a sign of respect is considered culture. But also the use of 5 year plans for macro economical planning by the government is considered culture. Because of this broad applicability of the term culture, the theoretical basis of this research was extended with a second framework.

A model was used that integrates culture with the different environments in which complex multi-actor problems are dealt with. This integrating model is the four layer model of Williamson on new institutional economics (Williamson 1998, 2000). See Fig. 13.1.

The four layer model of Williamson is briefly discussed here based on his work (Williamson 1998, 2000) and the interpretation of (Meijer 2009).

The model consists of four layers of social analysis, each with its own time scale which gives an indication of the pace of change in that level. At the first level informal institutions are listed like customs, traditions, norms and religion. These informal institutions change very slowly with a frequency estimated in terms of centuries. Level 2 incorporates the institutional environment. This includes the formal rules within society like laws. Level 3 is called governance and is about how different entities interact given the institutional environment. This includes the different types of contracting. At the fourth level the functioning of the firm itself is optimized by means of resource allocation and employment. This is a continuous process.

The arrows connecting the different levels indicate that the higher levels influence the lower ones. For example the informal institutions from level 1 influenced the formation of the laws in level 2. But the institutional environment in level 2 is not completely determined by the informal institutions in level 1. Parts of the institutional environment are consciously designed by going beyond taboos, customs, traditions, and codes of conduct. This structure of influence and design also applies to the lower levels of social analysis in the framework.

As stated, this four layer model integrates culture with the different environments in which complex multi-actor problems are dealt with. Applying the model to the complex multi-actor problem of the Indian electricity challenge provides the following example elements from the environments in which the electricity challenge needs to be solved.

---

[2]The individual differences in the group test are left out of the scope of this research as they can be explained by either personality or variation in the measurements.

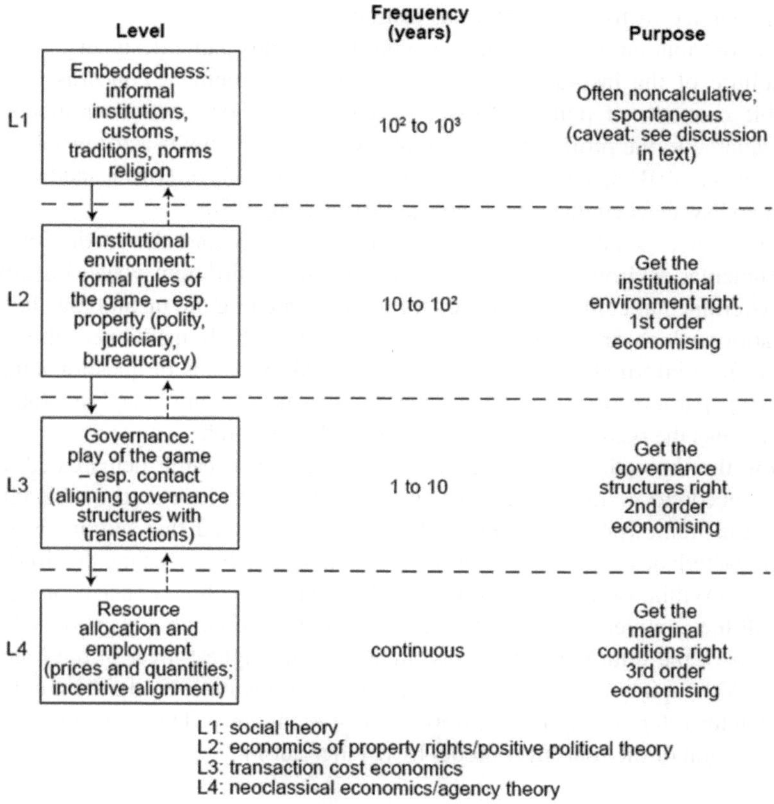

Fig. 13.1 Four layer model by (Williamson 2000)

- Level 1 – Informal institutions: Example of norms and values: From the perspective of respect, the average Indian will not contradict its superior.
- Level 2 – Institutional environment: Examples of the legislative structure of India: the Indian Electricity Act and the Energy Conservation Act.
- Level 3 – Governance environment: Examples of alignment of governance structures with transactions: Five year plans for macro economical planning and the New Hydro Policy.
- Level 4 – Resource allocation: Examples of the actual business: actual cost price electricity generation, specific subsidies for renewable electricity generation.

All these elements of the complex multi-actor environment are part of the culture that players bring to the game. As shown in the introduction of this chapter, complex multi-actor problems often stretch to multiple levels of the four layer model. But although all the levels are relevant, the choice was made to focus this research on the influence on games by the informal institutions situated in the first level of the model of Williamson. The influence of the institutional environment and the different governance structures are left out of the scope of this research.

This choice to focus on informal institutions was made since in this field the largest contribution can be made in supporting serious game designers. Assessing the culture of the institutional or governmental environment concerns the more tangible concepts of policies, laws and regulations. For these assessments tools are available to the professionals working with complex multi-actor problems like policy analysis (Bruin and Heuvelhof 2002), network analysis (Bruin and Heuvelhof 2002) and systems engineering (Sage and Armstrong 2000).

This choice is possible since informal institutions and the institutional and governmental environment are analyzed in complete different ways. Institutional and governmental culture is assessed through researching the institutions, laws and regulations which are in place by means of the methods mentioned above. The culture from informal institutions can be assessed by means of questionnaires and observing participants. This makes these assessments completely separate tasks which opens the possibility to focus on one in this research.

Now the theoretical basis was described and the research demarcated a final remark needs to be made regarding the term culture as it is used in this chapter. Using the framework of (Meijer 2009) it was explained that culture is one of the characteristics that players bring into the game. Next, using the four layer model of (Williamson 2000), it was described that this culture consists of elements from all the complex multi-actor environments. This research is demarcated to the influence of the informal institutions situated in the highest layer of the four layer model. Although culture is more than the informal institutions, the term culture in this chapter refers to the these informal institutions only. This is in line with the interpretation of the four layer model by (Meijer 2009).

13.3 Culture Driven Game Design Method

The Culture Driven Game Design Method was developed and tested in a collaboration project of the Delft University of Technology (TU Delft) in the Netherlands and the Centre of Science, Technology and Policy (CSTEP) from Bangalore, India. The Culture Driven Game Design Method consists of a three-step procedure that is to be inserted once in an iterative serious game design method. A schematic overview of the Culture Driven Game Design Method is given in Fig. 13.2.

The starting point for applying the Culture Driven Game Design Method is a near-final version of the game (Game version N in Fig. 13.2) which was developed through multiple iterations of playtesting using a team of playtesters. Note that these playtesters are not the targeted players, in most cases they are colleagues, friends or a testers panel employed by the game designer. Playtesting with these playtesters incurs (re)designing the game according playtest results until the player experience meets your criteria (Fullerton 2008). As a consequence of this repetitive playtesting with playtesters, chances are high that the game was adapted to the culture of this initial test group. Starting from this version of the game, our method proposes the following procedure:

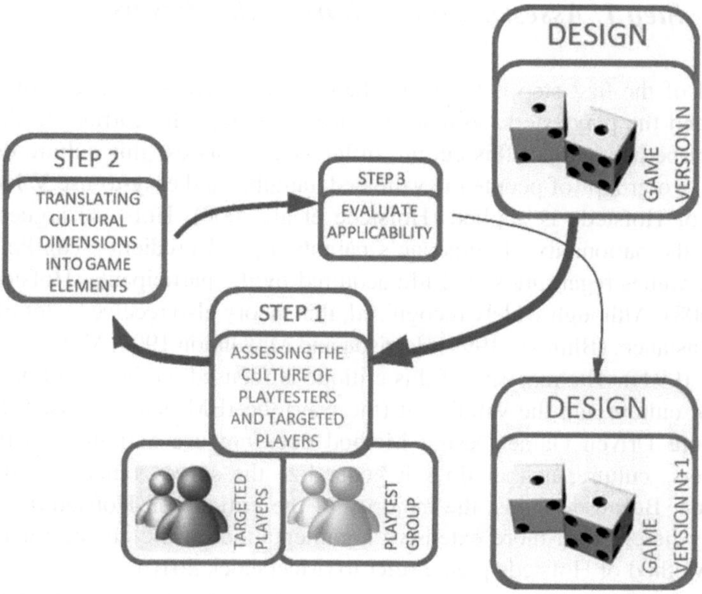

Fig. 13.2 The Culture Driven Game Design Method

- Step 1 – In order to adapt Game version N to the culture of the targeted players the difference in culture between the targeted players and the playtesters needs to be known. It is therefore that in step 1 the culture of both the targeted players and the playtesters is assessed. The output of this first step is a table with the culture assessment of both groups including the difference between them, presented on five culture dimensions.
- Step 2 – In the second step the culture dimensions are linked to game elements in a Cross Dimensional Matrix that we developed for this method. Using the matrix, a high culture difference on a culture dimension is linked to potential conflicts with game elements. For each potential conflict it is explained in the Culture Driven Game Design Method why there is a potential conflict and one or more suggestions are done to avoid or mitigate this conflict.
- Step 3 – However, before adapting the game using these suggestions the relevance of each potential conflict is determined. This is done by the game designer by interpreting the game in step 3. Once the relevance for all the potential conflicts indicated in step 2 is determined, it is up to the game designer to decide whether game elements should be removed, adjusted or kept in place.

The version following from the next design step (Game version N + 1 in Fig. 13.2) is adapted to the culture of the players. In the next three subsections each step is explained in more detail.

13.3.1 Step 1: Assessing the Culture of the Players

The goal of the first step is to assess the culture difference between the targeted players and the playtesters, so that in proceeding steps the current version of the game can be adapted for this culture difference. To assess this culture difference between two groups of people the validated, tangible and easy-to-use Value Survey Method of Hofstede is applied (Hofstede et al. 2008). Hofstede argues that by knowing the nationality of someone's parents a good prediction can be made of the basic values regarding social life acquired by the participants (Hofstede 1980, 2001, 2008). Although widely recognized, this theory also received a lot of critique (see for instance: (Bhimani 1999; Harrison and McKinnon 1999; McSweeney 2002; Redding 1994)). The majority of this critique is focused on the use of nations as a proxy for culture and the validity of (the way) the IBM data was used. However, the Culture Driven Game Design Method does not use nations as a proxy for the players' culture, neither does it generalize the survey outcomes to a larger population. Because of this, the majority of the critique on Hofstedes work does not apply here. For a more extensive argumentation on the choice for the Value Survey Method of Hofstede please refer to (Meershoek 2010).

The Value Survey Method assesses the culture difference between the targeted players and the playtesters by means of a questionnaire consisting of 20 questions (Hofstede 2008), structured along Hofstede's five culture dimensions. The dimensions are briefly introduced here whereas a more elaborate description can be found in (Hofstede 2001);

- *Power distance* – This dimension runs from egalitarian (small power distance) to hierarchical (large power distance) societies. It is the extent to which the less powerful members of organizations and institutions (like the family) accept and expect that power is distributed unequally.
- *Identity* – This dimension runs from collectivistic to individualistic societies. In individualistic societies a person is expected to look after himself or herself and his or her immediate family only. This in contrast with the collectivistic societies in which people from birth onwards are integrated into strong, cohesive in-groups, which continue to protect them throughout their lifetime in exchange for unquestioning loyalty.
- *Gender* – This dimension runs from feminine, 'sit and talk' societies to masculine, 'stand and fight' societies. Masculine societies have clearly distinct social gender roles; men have to be assertive, tough and focused on material success. Women are supposed to be more modest, tender and concerned with the quality of life. In feminine societies these social gender roles overlap: both women and men are supposed to be modest, tender and concerned with the quality of life.
- *Fear of the unknown* – This dimension opposes uncertainty-tolerant, novelty seeking cultures to uncertainty-avoiding, strangeness-fearing ones. In uncertainty avoiding cultures members of institutions and organizations within a society feel threatened by uncertain, unknown, ambiguous or unstructured situations.

Table 13.1 Example output of step 1 of the Culture Driven Game Design Method

	Players	Playtesters	Difference
Power Distance	19	4	15
Identity	6	96	90
Gender	53	–35	88
Fear For The Unknown	–62	–106	44
Gratification Of needs	38	41	3

– *Gratification of needs* – This dimension contrasts short-term oriented cultures to long-term oriented ones. A long term orientation stands for a society which values virtues oriented towards future rewards, in particular adaptation, perseverance and thrift. Short term orientation stands for a society which fosters virtues related to the past and present in particular respect for tradition, preservation of 'face' and fulfilling social obligations.

Using a spreadsheet, with the formulas given in (Hofstede et al. 2008), the output of the questionnaires can be processed. This results in the output of step 1; a table with the culture assessment of both the targeted players and the playtesters, including the difference between them, presented along Hofstede's five culture dimensions. An example of such output is shown in Table 13.1.

On the left side of Table 13.1 Hofstede's five culture dimensions are shown. In the first two columns the culture assessments are presented of the targeted players and playtesters respectively. The third column indicates the culture difference between these two groups. It is this culture difference that the game needs to be adapted for.

13.3.2 Step 2: Translating Culture Dimensions into Game Elements

In step 2 of the Culture Driven Game Design Method the culture difference on the culture dimensions is linked to the choices regarding game elements a game designer needs to make during the design process. By linking the culture difference to the choices regarding the game elements the method is able to actively support the game designer in adapting the game to the players' culture.

An overview of the choices a game designer needs to make is provided by the model of game dimensions by (Wenzler 2008). Wenzler first defined four basic components that each simulation game has. Each component is made up of four dimensions, representing the game structure. Each of these sixteen dimensions is then further defined into a range of possible states. The 16 game dimensions of Wenzler are crossed with Hofstede's five culture dimensions. This resulted in the Cross Dimensional Matrix depicted in Fig. 13.3.

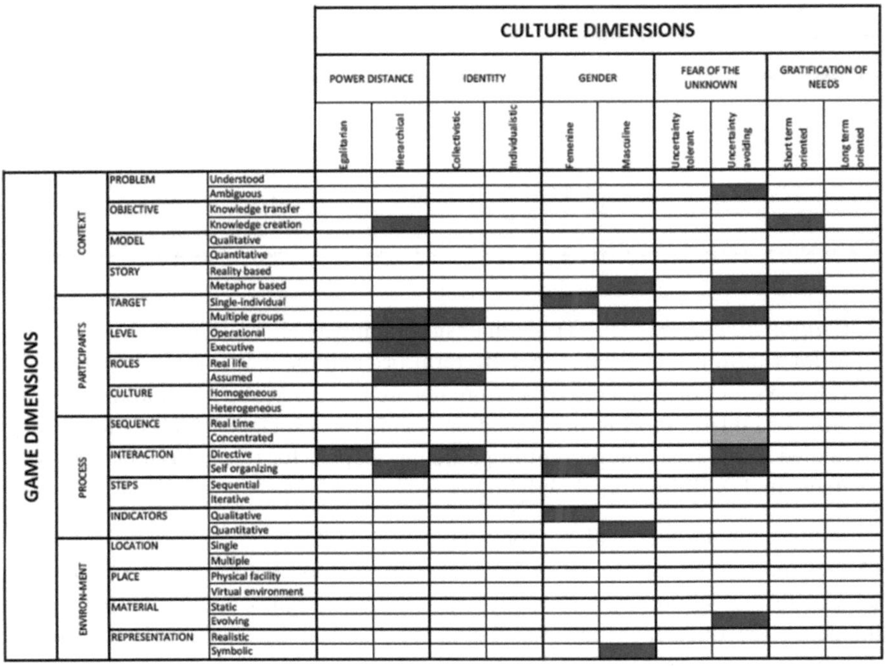

Fig. 13.3 The Cross Dimensional Matrix

In the Cross Dimensional Matrix the culture dimensions (x-axis) are crossed with the game dimensions (y-axis). Each cell of the Cross Dimensional Matrix represents a potential conflict between the players' culture[3] and a game element. So for each cell it was examined if the combination of the specific culture collides with the extreme on the game dimension. This examination resulted in a classification of each cell in three possible states:

- White – It cannot be deducted from theory neither is there an expectation that the combination of a high difference on the culture dimension and the extreme on the game dimension results in a potential conflict.
- Red – It can be deducted from theory that the combination of a high difference on the culture dimension and the extreme on the game dimension results in a potential conflict.
- Orange – Using a verifiable assumption it can be deducted from theory that the combination of a high difference on the culture dimension and the extreme on the game dimension results in a potential conflict.

[3] As stated in Sect. 13.2, this research is demarcated to the influence of informal institutions situated in the highest layer of the four layer model of Williamson. If other layers were to be included, the number of potential conflicts would increase as well as the amount and structure of the culture dimensions.

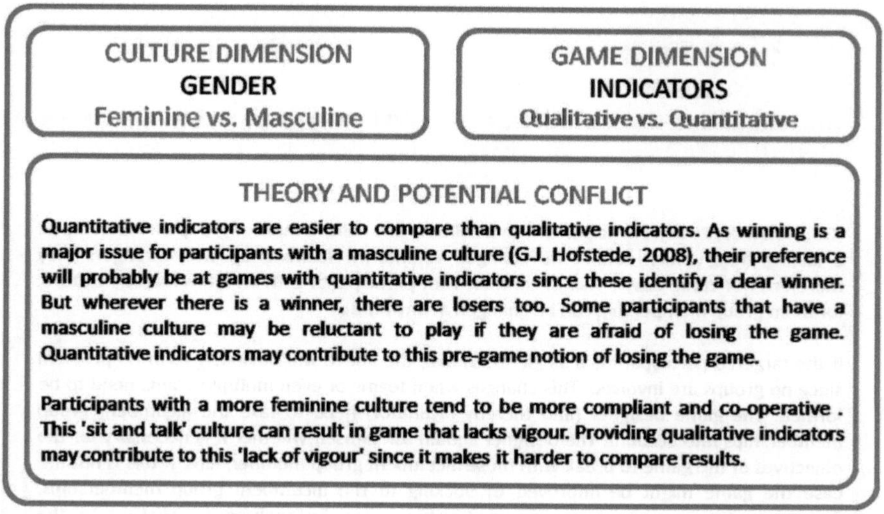

Fig. 13.4 Description of the potential conflict between Gender and Indicators

As shown in Fig. 13.3, 26 cells are classified as red. One cell is classified as orange and the other 293 cells are classified as white. The cells classified as red or orange are discussed in (Meershoek 2010).

Using the culture difference from the output of step 1, the serious game designer is able to track in the Cross Dimensional Matrix which choices regarding game elements potentially result in a conflict with the players' culture. Let us examine the example output of step 1 in Table 13.1. It concludes that there is a large culture difference between the targeted players and the playtesters on the identity and gender dimension. Using the Cross Dimensional Matrix in Fig. 13.3, ten potential conflicts are identified for these culture dimensions. Two of these potential conflicts are described here.

The first potential conflict results from the combination of high culture difference at the gender dimension and game elements situated at the indicators dimension. Figure 13.4 provides the description of this potential conflict.

The second potential conflict results from the combination of high culture difference at the identity dimension and game elements at the target dimension. Figure 13.5 provides the description of this potential conflict.

As shown in Figs. 13.4 and 13.5, a description of a potential conflict starts with the relevant dimensions. Next, an explanation is given of the ways in which the culture dimension affects the willingness to engage in gaming provided by (Hofstede 2008). This theory forms the final stepping stone towards the translation to the game dimension. In the translation the consequences of this willingness to engage in gaming for the specific game dimension are reasoned. The description is completed with a suggestion how to mitigate this potential conflict.

It is acknowledged that a relatively small amount of literature was available for providing the theory that forms the final stepping stone in the translation towards

CULTURE DIMENSION
IDENTITY
Individualistic vs. Collectivistic

GAME DIMENSION
TARGET
Single individual vs. Multiple groups

THEORY AND POTENTIAL CONFLICT

(G.J. Hofstede, 2008): Participants with a collectivistic culture think of group membership as a lasting thing which cannot be changed voluntarily. Teaming up with strangers may therefore not come easy . This in contrast with participants with an individualistic culture, for them group membership can be changed at any time .

If the targeted participant is a single individual, the above will obviously form no problem since no groups are involved. This changes when teams or even multiple teams need to be formed. The game designer (and thereby facilitator) need to take the incumbent group memberships into account. The designer should ask himself whether it is necessary for the objectives of the game to break with these incumbent group memberships. If this is not the case the game might be improved by sticking to the incumbent group memberships. Respecting the incumbent group memberships can be realized by sticking to the organizational structure or by leaving the team formation to the players.

Fig. 13.5 Description of the potential conflict between Identity and Target

game dimensions. The reason for this lacuna in the theory available is probably similar to the reason why the number of game design methods is limited. Serious game design is compared to other design sciences a young discipline (Mayer 2010; Salen and Zimmerman 2004). Further research by culture specialists in the field of the influence of culture in games is necessary to improve this translation from culture dimensions to game dimensions. This research may well provide the first step.

To complete step 2 of the Culture Driven Game Design Method the game designer transfers the culture difference between the targeted players and the playtesters to the Cross Dimensional Matrix. This provides the designer with an overview of all the potential conflicts. However, not all potential conflicts are relevant. This issue is addressed in the third step of the Culture Driven Game Design Method.

13.3.3 Step 3: Determining Relevance of Potential Conflicts

Not each potential conflict is relevant. Take for example a group of players who have a far more hierarchical culture than the playtesters used earlier in the design. Identified as possible conflict is the combination of the hierarchical culture with the mix of players from the operational and executive level. However, as stated in the description of this potential conflict, if the incumbent hierarchy is respected in the game, no conflict is to be expected. It is up to the game designers to determine the relevance of each conflict by interpreting their game.

Once the relevance for all the potential conflicts indicated in step 2 is determined, it is up to the game designer to decide whether game elements should be removed, adjusted or kept in place. Suggestions that can be used to mitigate the identified conflicts are stated in the description of the potential conflicts. However, game design remains a creative process which makes each game is different. This provides opportunities to the game designers to avoid the identified conflicts in their own manner. It is therefore that the Culture Driven Game Design Method identifies the potential conflicts between the current version of the game and the players' culture and provides suggestions on how to mitigate these conflicts. But it is up to the serious game designer to determine how to adapt the game to the players' culture.

The version that follows from the next design step is adapted to the culture of the targeted players. This is the next version of the game in the iterative game design process.

13.4 Evaluation

In the previous section the design of the Culture Driven Game Design Method was described. This section elaborates on the evaluation of this method. In the first subsection the set up of the evaluation is described. The second subsection provides a description of the game developed for this evaluation; the Indian Electricity Game (IEG). In the third subsection the results of the evaluation are presented. These results are discussed in the fourth subsection.

13.4.1 Evaluation Set Up

The Culture Driven Game Design Method was evaluated by comparing it to a benchmark method. As argued in the introduction; by means of playtesting with the targeted players it is possible to adapt a game to the culture of these players. It is therefore that the method of playtesting functioned as the benchmark method.

The Culture Driven Game Design Method was compared to this benchmark method in two elaborate case studies. Given the limited number of cases it makes sense to select cases which are the extreme opposites from each other (Eisenhardt 1989; Pettigrew 1990). For these cases expectations can be set which allows a better evaluation of the functioning of the Culture Driven Game Design Method. The following cases were prepared:

- Case Study 1 – In this case there was a large culture difference between the playtesters and the targeted players. The targeted players are employees of CSTEP.

- Case Study 2 – In this case there was no culture difference between the playtesters and the targeted players. The targeted players are students of the Delft University of Technology.

In these two case studies the versions of the IEG that resulted from the Culture Driven Game Design Method (IEG version C) were compared with the benchmark versions (IEG version PT) by means of a

- Static comparison – in which the versions were compared without playing them.
  During this comparison the game elements that were changed and the reasons why certain game elements were changed were examined.
- Dynamic comparison – in which the versions were compared by playing them.
  During this comparison the cultural fit during the sessions of both versions was examined.

Now the set up of the evaluation is described, the next subsection is dedicated to the game used in the evaluation; the Indian Electricity Game.

13.4.2 Evaluation Game: The Indian Electricity Game

The Indian Electricity Game was developed at the Centre for Study of Science, Technology, and Policy (CSTEP) in Bangalore, India. CSTEP is a private, non-profit organization with a vision to undertake research in engineering, science, and technology where it is relevant to India's economic and human development. CSTEP works in subjects such as energy, infrastructure, materials science, information and communications technologies, and security.

One of the challenges India faces which is relevant to both the economic and human development of the country is answering the increasing demand for electricity. This challenge has been the subject of various research projects by CSTEP and other organizations. In order to get relevant actors acquainted with the results of these research projects, the decision was taken to construct a game with this challenge as the subject. This game was named the Indian Electricity Game.

The IEG is a role-playing game for three persons that can be played in 2 h time. The three roles in the game are the Planning Commission, the Central Electricity Authority and the Ministry of New and Renewable Energy. These institutions play a central role in the planning of the extension of the electricity generation capacity. In India this planning is made using 5 year plans. In the IEG the players need to fill in two 5 year plans. This quantitative planning assignment forms the core of the game. When doing so they can choose from different generation plants using different energy sources like coal, gas, nuclear power, wind, hydroelectricity en solar PV. Each plant has different specifications regarding investment costs, generation costs and carbon emissions. The individual objectives attached to the different roles summon the players to minimize on these specifications. This forces the players to manage the trade-offs between the team objectives and their individual objectives during this planning process. In this way the players of the IEG gain insight in the various economical, technical, political and managerial issues that play a large role in the Indian electricity challenge. For a more extensive description of the IEG please turn to (Meershoek 2010).

13.4.3 Evaluation Results

In the previous subsection the set up of the evaluation was described. In this subsection the results of the evaluation are presented.

The static comparison in Case Study 1 showed that each of the 6 adjustments made in IEG version PT were also made or strived for in IEG version C. It was therefore concluded from the static comparison that the Culture Driven Game Design Method was just as able to adjust the IEG to the culture of the players as the playtest method. These results were confirmed in the dynamic comparison. In this dynamic comparison the targeted players played both versions. Independent observers determined the cultural fit of both versions by judging the questionnaires, interviews, and (videotaped) sessions. Both teams used similar strategies for both versions which resulted in a similar game process and game outcomes. The independent observers concluded that IEG version C had a slightly better cultural fit than IEG version PT. As IEG version PT was adjusted to the culture of the players by using the benchmark method of playtesting with the targeted players, it was concluded that the Culture Driven Game Design Method was able to adjust the IEG to the culture of the targeted players in Case Study 1.

In the second case study the results of the static and the dynamic comparison showed that the Culture Driven Game Design Method provided similar results as the playtest method. As the playtest method was the benchmark method, it was concluded that the Culture Driven Game Design Method was able to adjust the IEG to the culture of the players in Case Study 2.

Combining the results of the two case studies it was concluded that the Culture Driven Game Design Method was able to adapt the Indian Electricity Game to multiple groups of players with a different culture without playtesting the game with these players.

In order to generalize from a single case study Kennedy argued that one is to leave this to 'those individuals who wish to apply the evaluation findings to their own situations' (Kennedy 1979). To provide the possibility to those individuals, the case study and its context need to be described in detailed characteristics. It is then by the judgement of those individuals whether their situation is sufficiently alike the case study conducted, to generalize the evaluation outcomes.

The case studies conducted in this research can be described by examining the three main elements of the case studies; the Indian Electricity Game, the players, and the facilitators.

The Indian Electricity Game
- Is analogue,
- Combines role play with board gaming,
- Constitutes multiple conflicting interests,
- Is about a challenge that includes technical complexity.

All the players
- Are relatively young (<30 years),
- Graduated at university,
- Graduated for technical or beta degree,
- Knew each other before the sessions commenced,
- Formed a culturally homogeneous group at all four levels as described in the model of (Williamson 2000) in Sect. 1.2.

The facilitators
- Are relatively young (<30 years),
- Graduated at university,
- Graduated for a technical or beta degree,
- Worked at the same organization as the players,
- Knew the players before the sessions commenced.

13.4.4 Discussion

In this final subsection before the conclusions are drawn, we would like to discuss our expectations regarding the robustness of the evaluation outcomes. We expect that the evaluation results will be similar when the method is applied to other games that have slightly different characteristics. However, we advise caution when applying our method to other games that differ from the IEG in the following ways: (1) When the method is applied to a situation in which the players do not know each other, or the facilitator, before the session commences or (2) when the players have a different cultural background on the lower three levels of the model of Williamson. A change in one of these player characteristics is expected to result in a change in player interaction. For instance, for a large share of players, playing a game with friends results in different behaviour than playing a game with strangers. This final point of discussion brings us to the conclusion of this chapter.

13.5 Conclusions

The players of serious games are culturally sensitive agents; by means of their interaction with the game and the other players they bring their own culture into the game. Practice showed that if the game is not aligned with the culture of the players, this can result in conflicting behaviour that hampers the players to reach the objectives of the game. To become a success it is therefore necessary that the design of the game architecture is adjusted to the culture of its players. By playtesting with the targeted players, game designers are able to adjust their serious games to the culture of the targeted players. However, due to a lack of time, high costs and the need for a good first impression, playtesting is not always possible.

This chapter proposes the new Culture Driven Game Design Method; a valuable addition to existing serious game design methods that supports designers in adapting their games to the culture of the targeted players. The Culture Driven Game Design Method consists of a three-step procedure that is to be inserted in an iterative serious game design method. Our method provides a tool to assess and represent the culture of the targeted players as well as a set of guidelines to process this assessment and avoid conflicts between the culture of the players and the architecture of the game.

From the evaluation it was concluded that the Culture Driven Game Design Method was able to adapt the Indian Electricity Game to multiple groups of players with a different culture without playtesting the game with these players. It is expected that these evaluation outcomes can be generalized to cases in which the game, the personal player characteristics, or the personal facilitator characteristics are different.

We would like to close this chapter by providing two suggestions for further research. The first suggestion regards the translation from culture dimensions to game dimensions as described in the 27 conflicts in Sect. 13.3.2. It is acknowledged that a relatively small amount of literature was available for providing the theory that forms the final stepping stone in this translation. To improve the translation further research by culture specialists in the field of the influence of culture in games is necessary. This research may well provide the first step.

The second suggestion for further research relates to the focus of the Culture Driven Game Design Method. As discussed in Sect. 13.2, the method is focused on the influence on games by the informal institutions situated in the first level of the model of Williamson despite the relevance of the other levels. It would be interesting to examine if the available assessment tools for these lower levels can be integrated into the Culture Driven Game Design Method. This will create a comprehensive, tangible, and easy-to-use method that is able to actively support the game designer in adapting the game to the players' culture at any level this culture is expressed.

Acknowledgement We would like to thank our reviewers for their constructive comments that helped to improve this manuscript.

References

Abt, C. C. (1970). *Serious Games*. New York: The Viking Press.

Bhimani, A. (1999). Mapping methodological frontiers in cross-national management control research. *Accounting, Organizations and Society, 24*(28), 413–428.

Bruin, d., H., & Heuvelhof, t., E. (2002). Policy analysis and decision making in a network: how to improve the quality of analysis and the impact of decision making. *Impact assessment and project appraisal, 20*(4), 232–242.

Caluwé, d., L., Hofstede, G. J., & Peters, V. (2008). *Why do games work?* Deventer: Kluwer.

Consalvo, M. (2009). There is no magic circle. *Games and Culture, 4*(4), 408–417.
Crawford, C. (1984). *The art of computer game design*. Berkeley: McGraw-Hill.
Duke, R. D. (1974). *Gaming: the future's language*. New York: Sage Publications.
Duke, R. D., & Geurts, J. L. A. (2004). *Policy games for strategic management*. Amsterdam: Dutch University Press.
Eisenhardt, K. M. (1989). Building Theories from case study research. *The Academy of Management Review, 14*(4), 532–550.
Fine, G. A. (1983). *Shared fantasy: role playing games as social worlds*. Chicago: University of Chicago Press.
Fullerton, T. (2008). *Game design workshop: a playcentric approach to creating innovative games*. San Fransisco: Morgan Kaufmann.
Harrison, G. L., & McKinnon, J. L. (1999). Cross-cultural research in management control systems design: a review of the current state. *Accounting, Organizations and Society, 24*(5), 483–507.
Harvey, A. (2006). The liminal magic circle: boundaries, frames and participation in pervasive mobile games. *Journal of the Mobile Digital Commons Network,1*.
Hofstede, G. H. (1980). *Culture's consequences: International differences in work related values*. Beverly Hills: Sage.
Hofstede, G. H. (2001). *Culture's consequences*. Thousand Oaks: Sage Publications. Hofstede, G. H. (2008). Values Survey Module 2008: Questionnaire.
Hofstede, G. H., Hofstede, G. J., Minkov, M., & Vinken, H. (2008). Values Survey Module 2008: Manual.
Hofstede, G. J. (2008). One game does not fit all cultures. In d. Caluwé, L., G. J. Hofstede & V. Peters (Eds.), *Why do games work?* Deventer: Kluwer.
Huizinga, J. (1955). *Homo Ludens: A study of the play element in culture*. Boston: Beacon Press.
Kennedy, M. M. (1979). Generalizing from single case studies. *Evaluation quarterly,* 3(4), 661–678.
Klabbers, J. H. G. (2008). *The magic circle: principles of gaming & simulation* (2nd ed.). Rotterdam: Sense Publishers.
Kortmann, R., & Harteveld, C. (2009). *Agile game development: lessons learned from software engineering*. Paper presented at the Learn to Game, Game to Learn; the 40th Conference ISAGA, Singapore.
Mayer, I. (2008). Gaming for policy analysis: learning about complex multi-actor systems. In d. Caluwé, L., G. J. Hofstede & V. Peters (Eds.), *Why do games work?* Deventer: Kluwer.
Mayer, I. (2009). The gaming of policy and the politics of gaming: a review. *Simulation & Gaming, 40*(6), 825–862.
Mayer, I. (2010). Interview 1. In C. J. Meershoek (Ed.). Bangalore.
Mayer, I., & Veeneman, W. (2002). *Games in a world of infrastructures: simulationgames for research, learning and intervention*. Delft: Eburon.
McSweeney, B. (2002). Hofstede's model of national cultural differences and their consequences: A triumph of faith – a failure of analysis. *Human Relations, 55*(1), 89–118.
Meershoek, C. J. (2010). *The Culture Driven Game Design Method: Adapting serious games to the players' culture*. Delft University of Technology, Delft.
Meijer, S. A. (2009). *The organization of transactions: studying supply networks using game simulation*. Wageningen University, Wageningen.
Meijer, S. A., Hofstede, G. J., Beers, G., & Omta, S. W. F. (2006). Trust and tracing game: learning about transactions and embeddedness in the trade network. *Journal of Production Planning and Control, 17*(6), 569–583.
Nath, R. (1988). Regional culture, entrepreneurship, and high technology development in India. In J. W. Weiss (Ed.), *Regional cultures, managerial behavior and entrepreneurship*. New York: Quorum.
Paras, B., & Bizzocchi, J. (2005). *Game, motivation, and effective learning: An integrated model for educational game design*. Paper presented at the International DiGRA conference, Vancouver.

Pettigrew, A. M. (1990). Longitudinal field research on change: Theory and Practice. *Organization Science, 1*(3), 267–292.

Redding, S. G. (1994). Comparative management theory: Jungle, zoo or fossil bed. *Organization Studies, 15*, 323–359.

Sage, A. P., & Armstrong, J. E. (2000). *Introduction to Systems Engineering*. New York: Wiley-Interscience.

Salen, K., & Zimmerman, E. (2004). *Rules of play: Game design fundamentals*. Cambridge: MIT Press.

Watson, R. T., Ho, T. H., & Raman, K. S. (1994). Culture: a fourth dimension of Group Support Systems. *Communications of the ACM, 37*(10), 44–55.

Wenzler, I. (2008). Is your simulation green or blue? In d. Caluwé, L., G. J. Hofstede & V. Peters (Eds.), *Why do games work?* Deventer: Kluwer.

Williamson, O. E. (1998). The institutions of governance. *The American Economic Review, 88*(2), 75–79.

Williamson, O. E. (2000). The New Institutional Economics: Taking stock, looking ahead. *Journal of Economic Literature, 38*(3), 595–613.

MIX
Papier aus verantwortungsvollen Quellen
Paper from responsible sources
FSC® C105338

If you have any concerns about our products,
you can contact us on
ProductSafety@springernature.com

In case Publisher is established outside the EU,
the EU authorized representative is:
Springer Nature Customer Service Center GmbH
Europaplatz 3, 69115 Heidelberg, Germany

Printed by Libri Plureos GmbH
in Hamburg, Germany